Mingfeng Huang

# Optimum stiffness design of wind-excited tall buildings

Mingfeng Huang

# Optimum stiffness design of wind-excited tall buildings

## Computational design optimization considering uncertainties in wind engineering

VDM Verlag Dr. Müller

## Impressum/Imprint (nur für Deutschland/ only for Germany)

Bibliografische Information der Deutschen Nationalbibliothek: Die Deutsche Nationalbibliothek verzeichnet diese Publikation in der Deutschen Nationalbibliografie; detaillierte bibliografische Daten sind im Internet über http://dnb.d-nb.de abrufbar.

Alle in diesem Buch genannten Marken und Produktnamen unterliegen warenzeichen-, marken- oder patentrechtlichem Schutz bzw. sind Warenzeichen oder eingetragene Warenzeichen der jeweiligen Inhaber. Die Wiedergabe von Marken, Produktnamen, Gebrauchsnamen, Handelsnamen, Warenbezeichnungen u.s.w. in diesem Werk berechtigt auch ohne besondere Kennzeichnung nicht zu der Annahme, dass solche Namen im Sinne der Warenzeichen- und Markenschutzgesetzgebung als frei zu betrachten wären und daher von jedermann benutzt werden dürften.

Coverbild: www.purestockx.com

Verlag: VDM Verlag Dr. Müller Aktiengesellschaft & Co. KG
Dudweiler Landstr. 99, 66123 Saarbrücken, Deutschland
Telefon +49 681 9100-698, Telefax +49 681 9100-988, Email: info@vdm-verlag.de
Zugl.: Hong Kong, Hong Kong University of Science and Technology, Ph.D. Dissertation, 2008

Herstellung in Deutschland:
Schaltungsdienst Lange o.H.G., Berlin
Books on Demand GmbH, Norderstedt
Reha GmbH, Saarbrücken
Amazon Distribution GmbH, Leipzig
ISBN: 978-3-639-16253-0

## Imprint (only for USA, GB)

Bibliographic information published by the Deutsche Nationalbibliothek: The Deutsche Nationalbibliothek lists this publication in the Deutsche Nationalbibliografie; detailed bibliographic data are available in the Internet at http://dnb.d-nb.de.

Any brand names and product names mentioned in this book are subject to trademark, brand or patent protection and are trademarks or registered trademarks of their respective holders. The use of brand names, product names, common names, trade names, product descriptions etc. even without a particular marking in this works is in no way to be construed to mean that such names may be regarded as unrestricted in respect of trademark and brand protection legislation and could thus be used by anyone.

Cover image: www.purestockx.com

Publisher:
VDM Verlag Dr. Müller Aktiengesellschaft & Co. KG
Dudweiler Landstr. 99, 66123 Saarbrücken, Germany
Phone +49 681 9100-698, Fax +49 681 9100-988, Email: info@vdm-publishing.com
Hong Kong, Hong Kong University of Science and Technology, Ph.D. Dissertation, 2008

Printed in the U.S.A.
Printed in the U.K. by (see last page)
ISBN: 978-3-639-16253-0

# Acknowledgement

First and foremost, I am deeply indebted to my supervisor, Prof. C. M. Chan in HKUST for his intensive support, endless patience and eager encouragement. It is my great privilege to have his inspiring and thoughtful direction and guidance throughout my research. From the inception of this work, Prof. Chan has always been generous with his time and efforts and provided eminently insightful and detailed comments on everything from the methodological design to the writing process. His faith in me as well as his wide knowledge and critical thinking, which has always brought me back on track at difficult times, has been a godsend. Prof. Chan has taught me innumerable lessons and provided insights into the academic research work, and his mentorship was paramount in sharing his well-rounded experience with me on my long-term career goals.

I would like to offer my sincerest thanks to professors: Prof. K. C. S. Kwok and Prof. L. S. Katafygiotis, for their time and efforts to read this manuscript and providing many valuable comments that improved the contents of the work, as well as for their constructive suggestion and comments through this research work. Prof. W.J. Lou in ZJU also gave her suggestion on the preparion of this monograph.

I have enjoyed the camaraderie of structural engineering research groupmates: Kin-Ming Wong, Jie Zhang, Chun-Fai Yiu, and Juhui Zhang. I have benefited much from each of them at various times and places: their insightful and critical comments and suggestions, their collaboration and assistance in study, as well as the enjoyable atmosphere built by them with heart and soul.

I am also very appreciative of the assistances from the wind tunnel staff: Dr. P.A. Hitchcock and Mr. K. T. Tse, for their valuable experimental wind tunnel data.

It would be difficult to acknowledge everyone who has in some way or another contributed to the research reported in this work. During the past three and a half years, I have received help and encouragement from many people in different ways. Moreover, I owe much to my parents and sister for always believing in me and encouraging me to achieve my goals. Without their encouragement and understanding, it would have been impossible for me to finish this work.

# Table of Contents

# Nomenclature

## Latin letters

| | |
|---|---|
| $\hat{a}$ | Peak resultant acceleration |
| $A_i$ | Axial cross sectional area of a structural member |
| $b$ | Given boundary level for a particular process |
| $B$ | The building width normal to the approaching wind direction |
| $B_i$ | Breadth of rectangular concrete frame element |
| $d^U$ | The allowable displacement or interstory drift ratio limit |
| $D_i$ | Depth dimension of rectangular concrete frame element |
| $D_f$ | The failure domain |
| $D_s$ | The safe domain |
| $D(\cdot)$ | Mathematical standard deviation operator |
| $e_{ij}$ | strain energy coefficient |
| $E$ | The axial elastic material modulus |
| $E(\cdot)$ | Mathematical expectation operator |
| $f$ | Frequency |
| $f_j$ | Modal frequency of a building structure |
| $f(\cdot)$ | A general function |
| | Probability density function |
| $F$ | Cumulative probability distribution function |
| $\mathbf{F}$ | External force vector |
| $g$ | Peak factor |
| $g_f$ | Davenport peak factor |
| $g_e$ | The equivalent peak factor |
| $g_j$ | Design constraints |
| | Performance limit-state functions |
| $g_p$ | The probabilistic peak factor |
| $g_w$ | The Weibull peak factor |
| $g_G$ | The Gamma peak factor |
| $G$ | The shear elastic material modulus |
| $G_j$ | Performance limit-state functions in the standard normal space |
| $h(t)$ | The unit impulse response function |
| $h_k(\mathbf{x})$ | The equality constraints |

| | |
|---|---|
| $H$ | The building height |
| $H_j$ | The mechanical admittance function for $j$-th modal vibration |
| $I$ | Moments of inertia of a cross-section |
| $\mathbf{J_{x\_u}}$ | Jacobian matrix of all first-order derivatives of a vector-valued function |
| $k_j$ | The $j$-th modal stiffness of a building |
| $\mathbf{K}$ | Stiffness matrix of a building system |
| $L(\cdot)$ | Lagrangian function |
| $m(\cdot)$ | Moment function |
| $m_j$ | The $j$-th modal mass of a building |
| $\mathbf{m}(\cdot)$ | Moment vector function |
| $\mathbf{M}$ | Mass matrix of a building system |
| $N$ | Number of crossing events |
| $p$ | A specific probability value |
| | Distribution parameters of random variables |
| $P\{\cdot\}$ | Probability of a given event |
| $q$ | Bandwidth parameter of a random process |
| $q_j$ | The $j$-th Modal displacement |
| $Q_j$ | The $j$-th Mode generalized force |
| $r_{jk}$ | Intermodal correlation coefficient |
| $\mathbf{R}^n$ | Vector space of $n$-dimension |
| $R(\cdot)$ | The correlation function of a random process |
| $S_M$ | Base moment response spectrum |
| $S_Q$ | Modal force spectrum |
| $t$ | Time parameter |
| $t_i$ | Thickness of concrete shear wall element |
| $T_b$ | The first-passage time |
| $T_R(\cdot)$ | The Rosenblatt transformation |
| $u$ | The fluctuating component of wind speed |
| | The modal wind speed in a Gumbel distribution |
| $u_n$ | The characteristic largest peak response value |
| $\mathbf{u}$ | Displacement response vector |
| | The standard normal vector |
| $\mathbf{u}^*$ | The most probable failure point |
| $u(\cdot)$ | The unit step function |

| $\bar{U}$ | The mean component of wind speed |
|---|---|
| $v$ | The value of wind speed |
| $v_b$ | The mean our-crossing rate (level-crossing rate) of a random process from the level $b$ |
| $V$ | The annual largest wind speed |
| $V_R$ | The design wind speed corresponding to a $R$-year return period |
| $\mathbf{x}$ | Random vector describing system uncertainties |
| $\mathbf{X}$ | X-component displacement response vector |
| $Y(t)$ | A random response process |
| $Y_m$ | Peak value of $Y(t)$ |
| $Y_n$ | Extreme peak value of $Y(t)$ |
| $\mathbf{Y}$ | Y-component displacement response vector |
| $z_i$ | Generic element sizing design variables |
| $z_j$ | $j$-th component of state space vector $\mathbf{Z}$ |
| $z_i^L$ | The lower element sizing bounds for a sizing variable |
| $z_i^U$ | The upper element sizing bounds for a sizing variable |
| $\mathbf{Z}$ | State space vector |

Greek letters

| $\alpha_j$ | Regression constant for the $j$-th modal force spetrum $S_Q$ |
|---|---|
| $\beta$ | Reliability index in the FORM |
| $\beta_j$ | Regression constant for the $j$-th modal force spetrum $S_Q$ Reliability index related to the occupant comfort performance function of the $j$-th modal vibration |
| $\beta_n$ | A dispersion measure of the distribution of extreme peak response |
| $\gamma$ | The Euler constant |
| $\delta(\cdot)$ | Dirac delta function |
| $\varepsilon$ | Bandwith parameter of a random process |
| $\phi$ | Mode shape |
| $\kappa$ | Shape parameter of the Weibull peak distribution |
| $\lambda_j$ | The Lagrangian multiplier for the $j$-th design constraint |
| $\lambda_m$ | $m$-th order spectral moments of a random process |

| | |
|---|---|
| $\mu$ | Mean value of a random variable or a stationary random process |
| $\xi_j$ | Modal damping ratio |
| $\rho$ | Scale parameter of the Weibull peak distribution |
| $\rho_{jk}$ | CQC combination factor |
| $\sigma$ | Standard deviation or root mean square (RMS) of a random variable or a stationary random process |
| $\tau$ | Time duration |
| $\Phi(\cdot)$ | Standard normal cumulative distribution function |
| $\Phi$ | Mode shape matrix of a building system |
| $\varphi$ | The joint action factor |
| $\varpi$ | Circular frequency |

Other mathematical operation

| | |
|---|---|
| $\mathrm{erf}(\cdot)$ | The error function |
| $\nabla$ | Gradient of a scalar function |
| $\|\cdot\|$ | Euclidean norm |
| $\bigcup\limits_{i=1}^{N}$ | Union of $N$ events |
| $\bigcap\limits_{i=1}^{N}$ | Intersection of $N$ events |
| $*$ | Complex conjugate operator |

Abbreviations

| | |
|---|---|
| 3D | Three-dimensional |
| ABL | Atmospheric boundary layer |
| CA | Combined approximation |
| CAARC | Commonwealth Advisory Aeronautical Research Council |
| CFD | Computational fluid dynamics |
| CMD | Computational molecular dynamics |
| CDF | Cumulative distribution function |
| CQC | Complete quadratic combination |

| | |
|---|---|
| ESWLs | Equivalent static wind loads |
| EPSD | Evolutionary power spectral density |
| FEM | Finite element method |
| FORM | First-order reliability method |
| FPK | Fokker-Planck-Kolmogorov equation |
| GA | Genetic algorithm |
| GC | Gaussian closure |
| HFFB | High-frequency force balance |
| HLRF | Hasofer-Lind-Rackwitz-Fiessler algorithm |
| KKT | Karush-Kuhn-Tucker necessary conditions |
| LCR | Level-crossing rate |
| MCS | Monte Carlo simulation |
| MDOF | Multi-degree-of-freedom |
| MM5 | The 5$^{th}$ generation of Mesoscale wind climate model |
| MP | Mathematical programming |
| MPEC | Mathematical programming with equilibrium constraint |
| MPFP | Most probable failure point |
| OC | Optimality criteria |
| ODEs | Ordinary differential equations |
| PBSD | Performance-based seismic design |
| PBWD | Performance-based wind resistant design |
| PDEs | Prtial differential equations |
| PDF | Probability density function |
| POT | Peak over threshold |
| PSD | Power spectral density |
| RBDO | Reliability-based design optimization |
| RMS | Root-mean-square |
| SDF | Single-degree-of-freedom |
| SL | Stochastic linearization |
| SMPSS | Synchronous multi-pressure scanning system |
| SORM | Second-order reliability method |
| SRSS | The square root of sum of square combination |
| XPSD | Cross power spectra density |

# List of Figures

# List of Tables

# CHAPTER 1  Introduction

## 1.1  Background

Tall buildings historically emerged with the development of stronger and lighter construction materials, such as wrought iron and subsequently steel, after the industrial revolution in the nineteenth century. For a dynamic, modern metropolitan city, such as New York, London, Tokyo or Hong Kong, where tall buildings have been an effective way to make use of valuable and limited land. Recent boom in high-rise construction is the continual expansion with this urban form.

The reasons for recent trends of constructing skyscrapers involve many aspects, from historical evolution to social development, from technology innovation to cultural recognition, and from economical achievements to civilization. Surely for cities such as Hong Kong and Tokyo, it is a consequence of their insular locations and the exorbitant value of a limited supply of land. In these cities people have become used to population densities almost unthinkable in the West. There has been a long demand for working and living in high-rise buildings. More importantly however, a towering skyscraper is the supreme architectural and corporate gesture. Its height makes it an instantly recognizable entity, it exerts its presence on the city through its defiance of nature, and from within, its views garner an impression of supremacy. Recently, there has been a shift towards the creation of genuinely distinctive, as well as hugely ambitious, more complexly shaped buildings. Of the 10 tallest buildings in the world eight are now in Asia. The tallest, Taipei 101, was the first skyscraper to break through the 500m height barrier. With its extraordinary height of 509m, Taipei 101 is a self consciously Asian structure. Its unusual form is inspired by pagodas, the ideal - and only - native Asian paradigm for this typology. Its shape is distinctive and original. It may not be aesthetically most pleasing but it has become a fine and recognizable symbol for Taiwan.

Shanghai's 421m Jin Mao Tower in an elegant shape heralds China's extraordinary entry into the height contest helped along of course by Hong Kong's sharp skyline. Shanghai World Financial Center of 492 m and Shanghai Center of 580 m are currently under construction. These two buildings are located at the skyscraper-studded Lujiazui area of Pudong District and constitute a super high-rise building cluster together with Shanghai Jin Mao Tower. It is true that the title of tallest building is about to pass from Asia to the Middle East with the construction of Burj Dubai Tower in 2008. At close to half a mile in height, this particular giant is being built with a new construction speed.

1

In some parts of the world, tall buildings are necessary to house growing urban populations and to accommodate more closely interrelated business activities. However, tall building structures are expensive. Tall buildings consume vast amounts of increasingly expensive energy to construct and maintain; they can be vulnerable to natural and human-made hazards. Such disadvantages have led to new challenges for the design of a new generation of modern tall structures. All tall buildings must be designed to be not only safe over its intended life and serviceable for its intended function, but also resource efficient, environment and people friendly.

For most tall buildings, their shape and orientation are mainly driven by architectural inspirations, functional requirements and site limitations. In some cases, however, wind engineering and structural engineering also play significant roles in determining the shape and structural form of the building. This can be particularly the case with supertall buildings where wind controls many aspects of the structural design. In order to reduce the base overturning moments in Taipei 101 tower, a number of building models of the prototype with various corner shapes were conducted in wind tunnels (Irwin 2006). The end result of examining a series of corner modifications was the cross section with stepped corners, which achieved a 25% reduction in the wind-induced base moment. To reduce wind-induced vibrations, a 600-tonne pendulum tuned mass damper was installed at the upper observatory levels. During the conceptual design stage of the Burj Dubai Tower, high frequency force balance studies indicated that wind-induced loads and responses could be reduced significantly by re-orienting the axes of the tower so as to align the most unfavorable aerodynamic directions with the wind directions where strong winds were least likely to occur. The whole tower was rotated through 120 degrees to achieve this. As the design evolved, a series of five force balance tests were undertaken at various stages, with the results being used in the next iterative design cycle (Irwin and Baker 2005).

Although the current design practice is capable of delivering feasible designs of tall buildings, but the final design achieved tends to be conservative and by no means optimal in terms of construction cost and serviceable performance. Furthermore, ensuring safety and reliability in the design require a deeper understanding into the risks of hazards threatening buildings. The difficulty of the design problem is also compounded by the inherent uncertainties presented in the environmental loads and in the structural system properties. Therefore, developing an automated design optimization technique to deliver the most cost efficient and reliable structural design while satisfying all specified ultimate safety, serviceability and habitability design performance criteria has appeared to be very challenging to the engineering community.

Computer-based design optimization has emerged as a promising design methodology, in which a design problem has been firstly formulated into a mathematical optimization model, and then a theoretically sound and numerically reliable algorithm has been developed to solve the optimal design problems. In the tall building design, the mathematical optimization model is mainly composed of two components. One is the formulation of design optimization problem, which consists of design variables, design objectives, and design constraints. The other is the structural analysis of building system, which is achieved by solving the equation of motions governing the building behavior. Much attention is firstly put to the development of a mathematical optimization model for tall building designs in this research. Numerical algorithms are devised based on the rational mathematical optimization model, which really captures the major design factors and realistically reflects the behavior of the physical building system and the characteristics of its external wind loading conditions. Finally, computational algorithms for design optimization must be properly implemented in a computer program to build a computer-based user friendly platform for practical applications.

## 1.2   Scope and objectives

The ultimate aim of this research is to develop a design optimization technique that automatically seeks the most cost efficient design solution while satisfying all specified ultimate safety, serviceability and habitability design performance criteria formulated as deterministic and probabilistic design constraints. The specific components of this study are as follows:

1.   To devise a coupled dynamic analysis procedure for assessing responses of general asymmetrical tall buildings with three-dimensional (3D) mode shapes under wind excitation.

2.   To integrate the wind tunnel derived aerodynamic load analysis procedure into the stiffness design optimization method.

3.   To formulate and solve the dynamic serviceability design optimization problem of tall buildings subject to static drift and dynamic acceleration performance constraints.

4.   To investigate the time-variant reliability of wind-excited building structures using extreme value statistical analysis.

5.   To develop the dynamic response optimization technique in the time domain by properly treating time dependent performance constraints and time-variant probability constraints.

6.   To identify and model the major uncertainties involved in wind loading conditions and structural systems for assessing the reliability of tall buildings against wind-induced motion.

7.   To develop a reliability performance-based optimal design framework to solve the design optimization problems of wind-sensitive tall buildings subjected to both deterministic drift

3

and probabilistic acceleration performance constraints.

8. To validate the effectiveness of the computer-based optimal design technique through a wide range of practical building examples.

The outcome of this research will be an advanced computer-based optimal design technique that has a direct impact and benefit on structural optimization of tall buildings. With such an optimal design tool, structural engineers can afford to implement modern performance-based design concept and achieve a more economical and reliable design solution. The specific contributions of this research are outlined as follows:

1. Advance the state-of-the-art research in the challenging field of reliability performance based design optimization of tall buildings.

2. Provide a powerful optimal design tool capable of delivering cost-effective and reliable design of tall buildings under wind excitation.

3. Enhance the competitiveness and efficiency of the tall building design industry through implementation of the reliability performance based design mathematical model into an operational generic computer model or program.

## 1.3    Major challenges of the research

Safety, serviceability and reliability in design of tall buildings in a typhoon-prone area require an in-depth study into the dynamic properties of tall buildings, the risks of wind hazards, and wind effects on buildings. Economically sizing the structural system of a modern tall building to satisfy all performance design requirements is generally a very challenging task because such a system consisting of thousands of structural elements usually is very complex in nature. Even with the availability of today's finite element analysis software, the search for the optimum structural system for a tall building satisfying a multitude of performance design criteria is a rather difficult and laborious task. The difficulty of the problem is further compounded by the inherent uncertainties presented in wind loads and in building systems.

The primary research challenges of the work are described as follows:

1. Develop a comprehensive and accurate mathematical model for design optimization

The structural optimization problem to be addressed is involved with a variety of design variables, and different static and dynamic, deterministic and probabilistic design constraints. The design

4

objectives involved in the optimal performance-based structural design problem have also been defined in terms of different performance requirements against various severity levels of wind hazards. The explicit formulation of dynamic acceleration performance design constraints of a wind-excited building is quite challenging, since the prediction of wind-induced responses of contemporary complex tall buildings is more difficult than that of regular buildings with simple geometric shapes and one-dimensional (1D) mode shapes.

## 2. Dynamic serviceability and occupant comfort design

The two major performance indexes for wind-induced serviceability design of tall buildings are the lateral drift and acceleration responses. Modern high-rise buildings designed to satisfy static lateral drift criteria may still oscillate excessively during wind storms (Griffis 1993; Kareem 1992). The serviceability design for motion perception can be carried out by checking the magnitude of wind-induced vibration against the acceptability threshold of motion (or the so called, occupant comfort criteria). With increasing height, often accompanied by increased flexibility and low damping, super-tall building structures are more susceptible to wind-induced vibration. An effective structural design optimization approach for suppressing wind-induced motions in tall buildings is still lacking.

## 3. Dynamic response optimization

The dynamic response optimization methodology has been developed to account for the stochastic nature of wind-induced movements of tall buildings. However, other than the simulation-based methods, there is still no effective and efficient analytical methods for system reliability analysis of wind-excited building structures. The time-variant reliability analysis procedure is to be developed firstly and then integrated into the dynamic response optimization method to deal with probabilistic drift design constraints. The need for incorporating numerous dynamic constraints imposed at all instants in the time domain is a major difficulty in the development of dynamic response optimization technique.

## 4. Reliability-based structural optimization algorithms

Although a number of numerical algorithms are available for solving deterministic structural optimization problems, an efficient and robust approach for reliability-based design optimization (RBDO) of large-scale structures is still lacking. The original form of the reliability performance-based design optimization problem is a nested optimization problem. Any change in the design variables may require for a reevaluation of reliability of the dynamic serviceability

5

performance, in which reliability analysis itself is a computationally intensive numerical procedure. The enormous computational effort due to the nesting of optimization and reliability analysis makes the conventional RBDO method impractical for large-scale problems. Therefore, it is a challenge to develop the reliability performance-based design optimization technique for practical tall building designs against wind hazards.

## 1.4 Work organization

This work is divided into nine chapters. It begins with an introduction, Chapter 1. Chapter 2 reviews the concept and technique of performance-based engineering and design optimization, and the relevant technique in wind engineering. The fundamental principles of performance-based engineering and design optimization, the uncertainties involved in wind engineering, the dynamic response analysis method and reliability analysis method are reviewed and summarized in this chapter. Moreover, the general framework for reliability performance-based design and its applicability are discussed in Chapter 2. The limitations of existing studies and their developments in overcoming these limitations are also addressed.

Chapter 3 first reviews the available methods for coupled dynamic analysis of building system and the advantage and disadvantage of these methods. Semi-analytical methods in the frequency domain based on wind tunnel tests for predicting wind-induced responses of tall buildings are presented. Finally, the intermodal cross-correlation reflecting the statistical coupling between modal responses under spatiotemporally varying dynamic wind excitation has been investigated in detail.

Chapters 4 presents the analysis method of aerodynamic wind loading on tall buildings and the integration between the wind load analysis method and lateral drift design optimization of tall buildings. The integrated stiffness optimization technique is based on the rigorously derived Optimality Criteria (OC) method, which is to be developed to achieve the optimal distribution of element stiffness of the structural system satisfying the wind-induced drift design constraints. The integrated wind load analysis and stiffness design optimization is capable of achieving more cost-efficient structures when compared to the results using the traditional static drift optimization method without wind load updating. The additional benefits of wind-induced structural load reduction can also be achieved by the integrated stiffness optimization method.

Chapter 5 presents an integrated wind-induced dynamic analysis and computer-based design optimization technique for minimizing the structural cost of general tall buildings subject to static and dynamic serviceability design criteria. Once the wind-induced dynamic acceleration response of

6

a tall building structure is accurately determined and the acceleration design constraint is explicitly formulated, the Optimality Criteria (OC) method, which is presented in Chapter 4, has been further developed to solve the optimal dynamic serviceability design problem subject to both peak resultant acceleration criteria and frequency dependent modal acceleration criteria. Not only is the computer-based optimization technique capable of achieving the most economical distribution of element stiffness of practical tall building structures while satisfying multiple static drift and dynamic acceleration serviceability design requirements, but also the numerical optimal design method is computationally efficient since the final optimal design can generally be obtained in only a few number of reanalysis and redesign cycles.

Chapter 6 focuses on the time-variant reliability analysis of building structures under random wind excitation. The important properties of random response process, such as mean level-crossing rate and peak distribution, have been reinvestigated by using the statistical counting approach. Based on statistical analysis of peak responses and the statistics of extremes, the uncertainty in the prediction of the largest peak responses due to inherent variability in the wind-induced random vibrations is investigated in terms of peak factors, which is defined as a ratio of the largest peak response to the standard deviation response. Using new peak factors obtained in this study, various methods have been developed for predicting the expected maximum vibration response of a wind-excited tall building during finite time duration.

Chapter 7 formulates the reliability-based performance design optimization problem mathematically in terms of deterministic and random design variables. The design optimization problem has been formulated with time dependent probabilistic performance constraints. Various ways for treating numerous time dependent constraints in the discretized time domains are discussed. For simplification, the multiple outputs of a linear stiffness system can be sufficiently described by a scalar stochastic response process at a critical location from the structural design point of view. Therefore, the peak factor approach developed in Chapter 6 can be applied to estimate the expected maximum response and in turn the time-variant reliability.

Chapter 8 develops a computer-based reliability performance-based structural optimization technique for the dynamic serviceability design of wind-sensitive tall buildings. Major uncertainties involved in both the wind loadings and the structural systems are taken into account in the framework. The original coupled two-loop reliability-based design optimization problem is reformulated and decoupled into two separated sub problems, i.e., the inverse reliability problem and the deterministic stiffness design optimization problem. These two sub problems are then successfully solved using the inverse reliability method and Optimality Criteria (OC) method, respectively.

Chapter 9 summarizes the major contributions and findings of this research study. Possible extension of this study for future research is recommended.

# CHAPTER 2   Literature Review

## 2.1   Performance-based engineering and design

### 2.1.1   Performance-based seismic design (PBSD)

Historically, building codes have required that buildings be built to meet a minimum level of safety requirement. In current design practice, the code-based design is performed to search for a feasible design solution satisfying all prescriptive mostly empirical code specifications. However, buildings conforming to those codes have been seen a dramatic rise in earthquake related losses. As estimated by the Federal Emergency Management Agency (FEMA) in 2000, in the past ten years total earthquake related losses were twenty times larger than in the previous 30 years combined (FEMA 349) as shown in Figure 2.1. FEMA's expenditures related to earthquake losses have become an increasing percentage of its disaster assistance budget. Predictions are that future single earthquakes, which will inevitably occur, may result in losses of $50-100 billion each. Developers, Stakeholders and occupants have become painfully aware of the financial and social consequences of earthquakes and are demanding that practical and cost-effective means be developed to address the issues of damage control and loss reduction.

Over the past several decades, leading structural engineers have promoted the development and application of performance based seismic design (PBSD) concepts. The basic concept of PBSD is to provide engineers with the capability to design buildings that have a predictable and reliable performance in earthquake. In 1992, FEMA sponsored the development of national consensus guidelines for the seismic retrofit of buildings, the Applied Technology Council (ATC-33) project. This was the first attempt to standardize the performance-based approach. After that, the FEMA-273 (1997), National Earthquake Hazards Reduction Program (NEHRP) Guidelines for the seismic rehabilitation of buildings, provides nation-wide acceptable guidelines for the seismic rehabilitation of buildings in the United States. One of significant new features is that the document identifies the methods and design criteria to achieve different levels and ranges of seismic performance. The four building performance levels are collapse prevention, life safety, immediate occupancy and operational. The similar approach to define building performance levels also can be found in the modern seismic design codes (e.g., the Chinese Code for Seismic Design of Buildings GB50011-2001).

9

More recently, the performance-based seismic design and assessment guidelines for new steel moment frame buildings have been proposed by the SAC FEMA program (FEMA-350 2000). The proposed reliability-based, performance-oriented approach is based on realizing a performance objective expressed as the probability of exceeding a specified performance level (Cornel et al. 2002). Kim and Foutch (2007) made an attempt to extend application of the FEMA approach to RC shear wall buildings. In that study, such as the analytical model, damage measure, and definitions of collapses were examined, and the confidence levels of the buildings were determined utilizing the new model and parameters. It was shown that the FEMA approach can be also useful for RC shear wall buildings.

## 2.1.2    Performance-based wind resistant design (PBWD)

PBSD employs the concept of performance objectives. A performance objective is the specification of an acceptable level of damage to a building if it experiences an earthquake of a given severity. This creates a "sliding scale" whereby a building can be designed to perform in a manner that meets the owner's various economic and safety goals. Further, PBSD permits owners and other stakeholders to quantify financially or otherwise the expected risks to their buildings and to select a level of performance that meets their needs while maintaining a basic level of safety.

It is obvious that the concept of PBSD is generally applicable to building design against other kinds of nature or human-made hazards. For instance, wind loading is the major factor governing the structural design of most tall buildings in Hong Kong, situated in a typhoon-prone but low seismic region. As modern buildings get taller and increasingly slender, the effects of wind-induced motions become more pronounced and the amount of structural materials required for lateral and torsional resistance and serviceability occupant comfort increases drastically (Smith and Coull 1991). Although wind-induced performance levels (limit states) of buildings have not been well defined compared with seismic engineering, some research work towards wind-induced performance levels in associated with the performance design objectives (design criteria) can be found (Melbourne and Palmer 1992; Chock et al. 1998; Jain et al. 2001).

Wind-induced performance could be related to the system behavior of a building under wind loading. The system behavior is the essential knowledge to deliver a feasible engineering design solution. Wind-induced responses in terms of internal force, deflection, drift and acceleration, are comprehensively used to describe the system response behavior under wind loading. It is clear that performance levels are closely connected to the concept of limit state design, which has been widely adopted in most of modern design codes. Actually, the concept of performance levels represent an evolution of prescriptive rules for limit state designs that have changed during years as more is learned about building behavior. Furthermore, a performance based design option will facilitate design of buildings to higher standards without explicit specification in the code and will allow rapid implementation of innovative technology. When performance levels are tied to probable losses and risk issues in a reliability-based design framework, the owner's long-term capital planning strategies can be taken into account in the building design process.

Although it has been realized that wind-induced dynamic serviceability are generally the major concerns in the design of tall buildings, research in structural design and optimization for dynamic serviceability has not received as much as attention as that for ultimate strength limit state design of such structures. In fact, the design of tall buildings in windy climates is generally dominated by serviceability considerations in terms of wind-induced deflections and vibrations, rather than by member strength requirements (Griffis 1993). The deflections or deformations from all load types should not impair the strength or effective functioning of a structure, supporting elements or its components, nor cause damage to the finishes. For typical structures, the deflection limits are recommended by various building design codes. The first step in establishing a serviceability design criterion is to define the load under which it is to be checked. Wind loading criteria for both strength limit states and serviceability limit states in the United States are normally the same and typically are based on a 50-year return period wind for normal buildings and a 100-year recurrence interval wind for important structures. The design wind pressures given in Hong Kong wind code (HKCOP 2004) have been determined from the hourly mean wind speed and peak gust wind speed having a return period of 50 years.

Vibration and oscillation of a structure should be limited to avoid discomfort to users and damage contents. For special structures, including long-span bridges, large stadium roofs and chimneys, and wind sensitive tall buildings, wind tunnel model tests are recommended for their wind resistance design to meet various serviceability limits. The serviceability limit states on oscillation, deflection and acceleration should be checked to ensure serviceable condition for the structure. The checking procedure for motion perception can be performed by comparing the magnitude of wind-induced vibration and acceptability threshold of motion (or so call, occupant comfort criteria). It has been widely accepted that the perception of wind-induced motion is closely related to the acceleration response of buildings (Kwok et al. 2007). Both peak acceleration and standard deviation acceleration under extreme wind conditions of 10-year or 5-year return period are commonly used to represent building motion (Burton et al. 2007).

Several researchers have suggested introducing acceleration limit states in preventing occupant

11

discomfort as a new design performance level for wind-sensitive buildings (Griffis 1993; Isyumov 1994). Acceleration and occupant comfort criteria for buildings undergoing complex motions were discussed by Melbourne and Palmer (1992). In that study, acceleration criteria to achieve acceptable occupancy comfort in buildings have been developed in terms of peak accelerations as a function of motion frequency and return period. In terms of performance based wind engineering, Chock et al. (1998) proposed a compatible set of wind design criteria which would be "risk-consistent" with the current framework for performance based seismic engineering. Performance-based wind engineering design levels recommended by Chock et al are given in Table 2.1. In the table, the design life of a building, which represents the exposure time for the building, is typically taken as 50 years to evaluate the probability of exceedance from the particular wind design level.

In terms of wind climatology, it is necessary to performing a site-specific design wind speed investigation, in which wind directionality and topographic effects would be considered. Jain et al. (2001) proposed a probability based methodology used to determine site-specific performance based design wind speeds for use in wind tunnel measurements and building design. In that paper, it was demonstrated that using such site-specific extreme wind loads can often lead to cost savings for the building owner.

### 2.1.3   Prediction of typhoon risk

Among the natural sources of risk, earthquake and typhoon are undoubtedly the most prominent ones. The concept of typhoon risk for tall buildings includes natural factors of atmosphere and water global circulation, local climatological and topographical conditions in the area under consideration, the level of lateral stability of the building structures, constructions, the extent of damage from possible typhoon and their consequences, social and economic factors.

While a typhoon refers to a tropical cyclone occurring in the western Pacific or Indian oceans, a hurricane is a severe tropical cyclone originating in the equatorial regions of the Atlantic Ocean or Caribbean Sea, traveling north, northwest, or northeast from its point of origin, and usually involving heavy rains. Eastern China is suffering by typhoon attacks in an annual basis. For example, according to the report of Hong Kong Observatory, near 4 tropical cyclones with equal or greater than No. 8 Signals (with a sustained wind speed of 63-117 km/h) have attacked the Hong Kong area for each year during the years of 1956 to 2006. Hurricane is the major nature disaster threatening the east coast of North America. In recent years, before 2006 season, a doubling of hurricane activity for the North Atlantic basin has occurred as well as an increase in the number of storms that are hitting land (Goldenberg et al. 2001).

The increase in hurricane frequency and intensity in the North Atlantic basin may be attributed to global warming (McCullough and Kareem, 2007). Climate change will become an important factor to spark extreme weather including fierce hurricanes or typhoons. Climate scientists warned of a future characterised by extreme weather events - long and intense droughts, fierce hurricanes or typhoons, heatwaves, and rising sea levels - as a result of rising temperatures in the Intergovernmental Panel on Climate Change (IPCC) report 2001. Scientists now understand much more about how weather systems work, so are able to predict how temperature changes affect rainfall patterns and storms. Jonathan Overpeck, professor of atmospheric sciences at the University of Arizona and a lead author of the IPCC report, pointed out that due to global warming more intense hurricanes will occur accompanied by rising sea levels, which may cause the sea to surge over coastal land. That will make coastal cities like New Orleans much harder to protect.

Wind hazard risk studies could quantify randomness of extreme wind events in terms of statistical distribution and parameters, and provide basis for wind-induced performance based design. Some wind hazard models (e.g., typhoon models) that accurately model the average wind field observations are proposed and available in the literature (Vickery et al. 1995a, 1995b, 2000). The numerical simulation of wind hazards (e.g., tropical storm, tropical cyclones) originated in the middle of 20th century with axisymmetric models, in which flow variations in the azimuthal directions are ignored (Ooyama 1969). Using these typhoon models, it is not difficult to simulate extreme wind events for a local typhoon-prone area, like Hong Kong. Actually, hourly mean wind velocity profile for design purpose according to Code of Practice on Wind Effects in Hong Kong 2004 is obtained based on observed and simulated directionally independent typhoon wind speed data.

Since directional extreme wind speed data is less than non-directional (all directional) wind speed data, it is necessary to develop sophisticated numerical simulation method to produce enough directional wind speed data for statistically estimating two-dimensional (speed and directionality) wind velocity climate. Various sophisticated three-dimensional models were proposed to study the more realistic tropical cyclones (Shapiro 1983; Krishnamurti 1989). Recently, a successful simulation of hurricane Andrew using the Fifth Generation PSU/NCAR Mesoscale Model (MM5) with a horizontal resolution of 6km was reported (Liu et al. 1997, 1999). Such models given tropical synoptic conditions could also be applied in practice for short range weather forecasting (Zhang et al. 2000). The use of these models or in the frame of Monte Carlo simulations as a numerical tool to study statistical characteristics of extreme wind events has been receiving increased attention (Vickery et al. 2000; Matsui et al. 2002; Jain et al. 2001). Using proper probability treatment calibrated by local meteorological data, directional and magnitude characteristic of probability distribution of extreme wind events (two-dimensional wind velocity climate) could be evaluated by typhoon simulation (Simiu and Heckert 1998; Matsui et al. 2002). The extreme value distribution analysis is to be conducted on each set of directional wind speed data to estimate the joint probability density function of wind speed and direction.

Since the past extreme wind events in Hong Kong were mainly tropical cyclones, the focus of the discussion will be placed on the simulation of tropical cyclones. Tropical cyclones can be described using Newton's second law augmented by conservation laws for mass, thermodynamic energy, and water vapor. As mentioned, one of the most rigorously developed wind climate models is MM5 model (Grell et al. 1994), which is open source and publicly available (http://www.mmm.ucar.edu/mm5/). The PSU/NCAR mesoscale model (MM5) is a limited-area, nonhydrostatic, terrain-following sigma-coordinate model designed to simulate or predict mesoscale and regional-scale atmospheric circulation. The similar models have been successfully used for hurricane wind-speed prediction and weather forecast. The MM5 model can be customized using terrain data of Hong Kong to simulate tropical storms which influence Hong Kong area (Clark et al. 1997). In addition, the model can be simplified for use in Monte Carlo simulations to produce enough hurricane wind-speed data (Vickery et al. 2000).

The MM5 model uses compressible Navier-Stokes equations with pressure perturbation and temperature as prognostic conditions to describe an atmospheric motion. The governing partial differential equations (PDEs) of an atmospheric motion can be set up in terms of the terrain coordinates $(x, y, \sigma)$. The computation domain including local terrain information can be discretized horizontally using B-grid staggering scheme and vertically using non-dimensional pressure sigma $(\sigma)$ levels, which means that the lower grid levels follow the terrain while the upper surface is flat. Then the governing equations can be numerically solved with the split-time-step approach where sound waves induced by compressible air flow are treated semi-implicitly on the shorter step (Grell et al. 1994).

Based on daily recorded hourly mean wind speed data during synoptic wind storms from the local observatory, the parent wind speed, the distribution of the complete population of wind speeds at a particular site, can be modeled with a Weibull distribution (Gomes and Vickery 1978). Such information is important to estimate of fatigue damage for which account must be taken of damage accumulation over a range of wind storms. However, for structural safety design, extreme wind speed rather than daily wind speed is desired for predicting maximum wind effects and then designing the proportional structure to resist extreme wind loading. Using crossing theory of random process, the design wind speed with a particular recurrence interval can be derived form the parent distribution information and the annual mean crossing rate of exceedances of a specified wind speed using Rice formula (1945), which is discussed in the later section about time-variant reliability. Then the return period associated with the exceedance of the specified design wind velocity is readily obtained as a reciprocal of the annual mean crossing rate.

Design wind speed could also be conveniently estimated by making use of the annual extreme wind speed values from the local observatory or numerical simulation, directly. In this approach, design wind speed could be estimated by statistical analysis to best fit observed or simulated typhoon wind speed data into the cumulative distribution function of Type I asymptotic extreme value distribution (Simiu and Scanlan 1996; Zuranski and Jaspinska 1996; Minciarelli et al. 2001). Kasperski 2007 systematically discussed the relationship between the design wind speed estimation and the design wind load codification. In that study, it is highlighted that the analysis of the extreme wind speed has to be based on separate storm mechanisms, e.g., strong frontal depression or other synoptic storms, gust fronts, thunderstorms and tropical cyclones.

### 2.1.4   Uncertainty modeling in wind engineering

For wind-excited buildings, major uncertainties rise from either aerodynamic wind loading characteristics or system properties. The uncertainty of wind loading characteristics is directly affected by statistical properties of design wind speeds, which could be quantified based on the probabilistic wind speed data analysis as reviewed in the previous section. System properties, including mass, stiffness and damping, are physically determined by the construction material, the structural member and form. These structural properties further influence the dynamic characteristics of structures, i.e., the natural frequency and vibration mode shapes. Various levels of practical randomness due to fabrication and construction process arising from the material, member or form may cause the structural system exhibiting uncertain behavior. In addition, the employ of computer models to predict system behavior inevitably introduces the model error due to the limits on resolution of the model or lack of sufficient knowledge and data.

A great extent of literature on uncertainties of the structural systems and wind characteristics is available in the context of wind engineering (Kareem 1987; Solari 1997; Hong et al. 2001). Kareem (1987) presented an overall probabilistic viewpoint for wind effects on structures. The discussion encompassed diverse scientific fields such as meteorology, fluid dynamics, statistical theory of turbulence, structural dynamics and probabilistic methods. Solari (1997) provided closed form expressions of the first and second statistical moments of the maximum response taking the uncertainties of the parameters and the idealized model error into account for wind-excited

15

structures. The probability distributions of wind-induced maximum responses were also studied in the parameter-space by Hong et al. (2001). Reliability analysis results in that paper suggest that the consideration of the uncertainty in structural parameters such as the fundamental frequency of vibration and the damping ratio is very important for the occupant comfort design, and that these uncertainties may be ignored for the strength-based or drift design. Solari (2002) discussed the prospects of analytical methods for estimating the wind-induced response of structures and quantifying uncertainties. Closed form solutions related to the 3D gust effect factor technique were extended to solve the complex probabilistic problems about the propagation of the parametric uncertainties and structural reliability.

The design wind speed is statistically estimated from the limited number of recorded or simulated data sample, which is inherently involved some uncertainties, i.e., sampling error, wind climate model variations. Peterka (1992) has shown that predictions of 50-year-retUren period wind speeds from record lengths of 20 to 40 years at a single station can have a significant uncertainty associated with sampling error, which is mainly caused by using the limited number of sample data from a short record duration. The uncertainty in the estimation of design wind speeds could be quantified by studying the mean and variance of the estimator for design wind speeds (Grigoriu 1982; Minciarelli et al. 2001). Grigoriu (1982) developed an estimator of the design wind speed from short records by taking into account the correlation between data points. Minciarelli et al. (2001) proposed a probabilistic methodology to reexamining wind load factors adopted in the standard provisions of ASCE 7-98. The study demonstrates that the sampling errors in the estimation of design wind speed and knowledge uncertainties have a significant influence on the determination of wind load factors conforming to a specified safety index.

Structural damping has remained the most uncertain parameter in the dynamic design of civil engineering structures. Structural damping prediction relies upon a high degree of empiricism that is almost entirely based upon full-scale measurements (Glanville et al. 1996). Several damping empirical models were derived form the analysis of damping data obtained from a wide and homogeneous set of full-scale measurements of buildings and other structures (Lagomarsino 1993; Lagomarsino and Pagnini 1995). A log-normal probability distribution function is usually accepted for estimating the statistical properties of the random structural damping (Solari 1996; Pagnini and Solari 1998). Due to lack of precise knowledge on damping mechanism, the prediction and selection of an appropriate damping value is still a subject of discussion and controversy in design practice (Kareem 1987). In general, lower damping in the fundamental mode is assumed since most building structures deform as a rigid body in the elastic range with very little portion of energy being dissipated by deformation of elements and structural connections. Tamura et al. (1994, 1996)

16

and Kitamura et al. (1995) introduced relevant advances in the knowledge of damping dependence on motion amplitude. Some research studies (Kareem and Gurley 1996; Kareem et al 1998) have also examined the values of damping in higher modes. It is believed that in higher modes a building generally experiences more deformation in elements leading to the result of higher damping.

## 2.2 Dynamic response analysis of wind-excited building systems

### 2.2.1 Analysis method in time domain

Modern tall buildings are wind sensitive structures. Recent trends towards developing increasingly taller and irregularly-shaped complex buildings imply that these structures are potentially more responsive to wind excitation. Making accurate predictions of wind loads and their effects on such structures is a necessary step in the design synthesis.

The response of structural systems to external excitations is a general and essential problem in structural dynamics. The procedure for obtaining the response of a structural system to external forces depends to a large extent on the characteristics of the structural system and the type of excitation. The classification of the dynamic analysis problems are given in Figure 2.2. In terms of deterministic or stochastic systems characterized by linear or nonlinear, the problems are classified into 4 sub-problems denoted by DL (deterministic linear system), DN (deterministic nonlinear system), SL (stochastic linear system), and SN (stochastic nonlinear system). Due to the principle of superposition, which clearly applies to linear systems alone, the responses of linear systems to a given number of distinct excitations can be obtained separately and then combined to obtain the total responses. It is because of this principle that the theory of linear systems is so well developed compared to that of nonlinear systems. Each sub-problem (DL, DN, SL and SN) can be further separated into two by consideration of external deterministic or random excitation. For example, the DL system has DLD (DL system subject to deterministic loading) and DLR (DL system subject to random excitation) problems.

The physical sources of wind loads on buildings are complex such that wind forces generated by windstorms can never be described or predicted perfectly (or deterministically). Generally speaking, wind loads on buildings are determined by the pressure distribution around the building structure, and in turn are controlled by the pattern of air flow around the structure. Since wind is a turbulent flow, the actual pressure distribution around building structures displays a complex pattern of variation in space-in two or three dimensions-as well as variation with time. When the degree of

17

disorder is sufficiently large, there is usually merit and economy in probabilistic rather than deterministic models. In this sense, wind-induced pressure and in turn wind-induced loadings could be treated as random field or random process. The term "field" indicates that the index space for random variables (pressures or loadings) is multi-dimensional. When a random field depends only on a single parameter, usually time, the random filed is reduced into a random process. Random process is a time indexed family of random variables $X(t)$, which could be observed at discrete points on a time axis.

Although some properties of building systems, e.g., damping, are stochastic in nature, building systems and their structural components are conveniently treated as deterministic in current design practice. Since wind-induced displacements of buildings are relative small compared with the dimensions of the building structures and their components, it is reasonable to assume the building systems are linear. Therefore, the dynamic response analysis for wind-excited buildings is a DLR problem as categorized in Figure 2.2. The DLR problem is a major subject in random vibration theory (Grandall 1963; Yang 1986).

Before moving to the probabilistic response of a system to a random excitation, it is necessary to study the time-domain method to predict the response of a linear system to an arbitrary but deterministic excitation. Due to the principle of superposition, a generic response $y(t)$ to a deterministic excitation $f(t)$ could be analytically obtained using the method of impulse response function (Meirovitch 1986)

$$y(t) = \int_0^t f(\tau) h(t-\tau) d\tau \qquad (2.1)$$

which is known as the convolution integral, and expressed the response as a superposition of the unit impulse response function, $h(t)$. It is noted that the impulse response function $h(t)$ in Eq. (2.1) is delayed, or shifted, by the time $t = \tau$. The convolution integral could be performed numerically and have been used in the investigation on the geometry of random vibrations (Der Kiureghian 2000) as well as the first excursion probabilities for linear systems by efficient importance sampling (Au and Beck 2001).

However, for a large-scale system, such as building systems, the method of impulse response functions becomes cumbersome. Moreover, the method is not applicable to nonlinear systems. The direct time-stepping method applying to linear or nonlinear equations of motion in three-dimensional (3D) space or in generalized modal space becomes popular with the increase of computer power. The time history analysis method has been implemented in the most of commercial Finite Element Method (FEM) software. The idea of time-stepping methods is to find the responses at next time step given the responses at previous step based upon finite time differences. The step-by-step Newmark method (Newmark 1959; Clough and Penzien, 1993; Chopra 2000) has been widely used in the numerical evaluation of dynamic responses for structural systems. As for any numerical computation procedure the accuracy of this step-by-step method will depend on the length of the time increment $\Delta t$. In general, the ratio of the time increment to the vibration period $T$ satisfying $\Delta t / T \leq 0.1$ will give reliable and stable numerical results (Chopra 2000).

The dynamic response of a structural system to a stochastic excitation is probabilistic in nature. The statistical characteristics of the random response process $Y(t)$ in second-order analysis can be effectively described by the correlation function $R_Y (t_1, t_2)$ at any two different time instant, which can be related to the correlation function of excitation $R_F (\tau_1, \tau_2)$ by

$$R_Y (t_1, t_2) = \int_0^{t_2} \int_0^{t_1} h(t_1 - \tau_1) h(t_2 - \tau_2) R_F (\tau_1, \tau_2) d\tau_1 d\tau_2 \qquad (2.2)$$

If both the input and the output are stationary stochastic process, it is expedient to change the variables to $t_2 - t_1 = \tau, t_1 - \tau_1 = \theta_1, t_2 - \tau_2 = \theta_2$; as a result, the above correlation function of response becomes

$$R_Y (\tau) = \int_0^\infty \int_0^\infty h(\theta_1) h(\theta_2) R_F (\tau + \theta_1 - \theta_2) d\theta_1 d\theta_2 \qquad (2.3)$$

It is noted that the correlation function approach is based on the method of impulse response functions. Such an approach is generally applicable to the response analysis problems for certain or uncertain linear systems (Katafygiotis and Beck 1995), e.g., DLR, SLD and SLR. The uncertainty characteristic could be reflected in the impulse response function, and the properties of random excitation could be described by $R_F (\tau_1, \tau_2)$. For instance, the integral on the right side of Eq. (2.3) can easily be calculated if the external excitation is a stationary delta-correlated process (white noise), which has the form of the following

$$R_F (\tau) = s \delta(\tau) \qquad (2.4)$$

where $\delta(t)$ is a Dirac delta function and $s$ is the intensity of the white noise. Such a correlation structure has been widely used for simplifying the random excitation process, which has a broader spectrum with wider frequency band covering the fundamental vibration frequency value of a structural system.

It is known that Gaussian processes are fully determined by their mathematical expectations and

19

correlation functions and that a Gaussian process remains Gaussian after passing through a linear system. This makes a correlation description a convenient technique for the analysis of random vibrations of a deterministic linear system, e.g., DLR. Furthermore, when the time interval $\tau$ separating the two measuring points is zero then the correlation function gives the mean square value for the random response process as

$$R_Y\left(\tau=0\right)=E\left[Y\left(t\right)^2\right] \tag{2.5}$$

where $E(.)$ denote the operator of mathematical expectation.

For instance, a generalized wind force process, practically obtained from linear combination of base moments or torque processes for a tall building excited by atmospheric turbulence, could be reasonably assumed as a Gaussian process due to the central limit theorem, which claims that the probability distribution of the sum of $M$ independent identically distribution random variables tends to become Gaussian when the number $M$ increases. The key requirements for applying the central limit theorem is that there must be an aggregation of many weakly correlated random effects and that no single effect (or small subset of effects) accounts for a dominant fraction of the total variance. This is the typical condition for building aerodynamic that turbulent flows with varying sizes and intensities do not follow the surface of the building body, but detach from it leaving regions of separated flow and a wide trailing wake or vortex, causing very irregular and complicated wind-induced pressure pattern acting on the building. Although the computational modeling of building aerodynamics becomes possible with the remarkable advances in Computational Fluid Dynamics (CFD) (Stathopoulos 1997 and 2002; Murakami 2002), physical modeling of building aerodynamics by using scale models in atmospheric boundary layer (ABL) wind tunnels is still a more reliable and practical way to quantify wind effects on building (Cermak 2003). In the wind-tunnel based semi-analytical approach, the Gaussian assumption for wind excitations and building responses are widely adopted to predict wind-induced responses of tall buildings.

Problems of nonlinear systems in the theory of random vibrations, e.g., DNR, SND and SNR in Figure 2.2, are much more difficult than those of linear systems because the principle of superposition is generally not applicable. Numerical method, such as the time-stepping method, becomes one necessary approach to investigate nonlinear response behavior. As an attempt to develop the theory of nonlinear systems, the method of moment or cumulant functions, based on the solution of differential equations for moment or cumulant functions, is extensively used to analytically investigate the problems of nonlinear statistical dynamics (Di Paola and Muscolino

1990; Di Paola et al. 1992; Papadimitriou and Lutes 1996; Papadimitriou et al. 1999). The method of moment functions is more conveniently described in a state space by introducing state variables or vectors as $\mathbf{Z}$, which could be related to the response processes $Y$ and their derivatives with respect to time as $\mathbf{Z} = \{Y, \dot{Y}\}$. Hence, various moment functions of the order $r$ about state vectors with $n$ components of $z_j$ could be notated as

$$m_{\underset{r}{jkl\ldots}}(t) = E\left[\underbrace{z_j(t)\,z_k(t)\,z_l(t)\cdots}_{r}\right] \quad j,k,l = 1,2,\cdots n \qquad (2.6)$$

A moment vector functions consisting of moments of the same order $r$ could be written as

$$\mathbf{m}_r(t) = \left\{ m_{\underset{r}{11\ldots}}(t),\; m_{\underset{r}{11\ldots2}}(t),\cdots \right\} \qquad (2.7)$$

Differential equations for $\mathbf{m}_r(t)$ can be obtained either by directly averaging the system's equations of motion (Papadimitriou and Lutes 1996), or by using Itô stochastic differential equation and by the operation of averaging (Papadimitriou et al. 1999), or by applying the appropriate Fokker-Planck-Kolmogorov (FPK) equation (Guo 1999; Paola and Sofi 2002; Li and Chen 2006). As a result, an equation whose left side there is a time derivative with respect to the moment function is obtained as the following

$$\frac{d\mathbf{m}_r}{dt} = \varphi_r(\mathbf{m}_1,\, \mathbf{m}_2,\cdots\mathbf{m}_r,\cdots) \quad (r = 1,2,\cdots ) \qquad (2.8)$$

The above hierarchy of equations represents a full description on the original stochastic dynamic problem. However, the system of equations is an infinite system. As a consequence, there arise the problem of the closure of the system in Eq. (2.8), i.e. of its reduction to a closed system of a finite number of equations. Suitable closure schemes have to be adopted. The simplest closure scheme is the Gaussian closure (GC) (Ibrahim et al. 1985) in which the higher order moments are expressed in terms of the first two moments, as if the response process were a Gaussian one. In the special case of purely additive white noise excitations, this procedure is analogous to another extensively used approximate method for predicting nonlinear responses, called stochastic linearization (SL) (Roberts and Spanos 1990). Though this procedure is simple and versatile, however, the Gaussian approximation is not satisfactory in the case of highly nonlinear systems and provides inaccurate results in terms of probability density function (PDF), so that the various response statistics needed for reliability analysis are not predicted adequately. Several other approximate techniques are available in the literature, such as the cumulant-neglect closure (Wu and Lin 1984), the stochastic averaging method (Roberts and Spanos 1986; Lin and Cai 2000), the dissipation energy balancing method (Cai and Lin 1988), the multi-Gaussian closure (Er 1998), the method of weighted residuals (Liu and Davies 1990).

As presented in above, numerous publications on randomly excited nonlinear dynamic systems have been appeared in the literature and many procedures for obtaining the exact and approximate solutions have been proposed (Zhu and Huang 2001). Other than the approximate method, such as the method of moment functions, the one for obtaining the exact stationary solutions is still attractive. For randomly excited nonlinear dynamic systems, the exact solutions are usually very difficult to obtain. Only when the excitations are Gaussian white noises, the exact solution is possible to obtain by solving the FPK equation governing the transition probability density of a Markov process, together with initial and boundary conditions (Caughey 1971; Caughey and Ma 1982; Dimentberg 1982; Zhu 1990). Recently, Proppe (2003) presented exact stationary probability density functions for nonlinear systems under Poisson white noise excitation. Dimentberg (2005) obtained an explicit expression of a stationary joint probability density of displacements and velocities, which is served as an exact analytical solution to the corresponding FPK equation for random vibrations of a rotating shaft with non-linear damping. Further developments in random vibration of nonlinear systems can certainly be expected in the future.

### 2.2.2 Analysis method in frequency domain

The Fourier transform provides the classical method for decomposing a time history into its frequency components. The Fourier transform of a realization of response process $y(t)$ in Eq. (2.1) could be obtained as,

$$Y(\omega) = H(\omega) F(\omega) \tag{2.9}$$

where $F(\omega)$ denotes the Fourier transform of the excitation; $H(\omega)$ denotes the frequency response function, and is obtained as the Fourier transform of the unit impulse response.

Due to Wiener-Khinchine relationship, the power spectral density (PSD) function of the stationary random response process $Y(t)$ could be related to the correlation function $R_Y(\tau)$ as

$$S_Y(\omega) = \frac{1}{2\pi} \int_{-\infty}^{\infty} R_Y(\tau) e^{-i\omega\tau} d\tau \tag{2.10}$$

On the other hand, the correlation function of the loading process $F(t)$ can be expressed as the inverse Fourier transform as

$$R_F(\tau) = \int_{-\infty}^{\infty} S_F(\omega) e^{i\omega\tau} d\omega \qquad (2.11)$$

Substituting Eq. (2.11) into Eq. (2.3), and then successively into (2.10), a simple algebraic expression relating the PSDs of the excitation and response could be obtained as (Meirovitch 1986)

$$S_Y(\omega) = |H(\omega)|^2 S_F(\omega) \qquad (2.12)$$

which is the fundamental frequency domain relationship regarded as the importance result of second-order analysis of the response of a linear dynamic system. It can be viewed as a frequency domain form of Eq. (2.3), which gives the second moment function of the stationary response. A major difference is that finding the response second moment or autocorrelation function from Eq. (2.3) involves a double integral in the time domain, while Eq. (2.12) simply gives us the response PSD from a multiplication of functions in the frequency domain.

Wind loads on structures consist of the mean component and the fluctuating component. For sufficiently rigid structure, the design wind loading is mainly associated with the pressure on the surface of the structure being, as a first approximation, proportional to the square of the wind speed. Wind speeds in the atmospheric boundary layer have generally been treated as stationary random processes, which can be represented as

$$U(t) = \bar{U} + u(t) \qquad (2.13)$$

where $\bar{U} =$ the mean component of the process; $u(t) =$ the fluctuating component with zero mean. The original fluctuating wind speed component $u(t)$ is essentially a summation of harmonic terms as

$$u(t) = \int_{-\infty}^{\infty} S_u(\omega) e^{i\omega t} d\omega \qquad (2.14)$$

where $S_u(\omega) =$ the PSD of fluctuating wind speed. Eq. (2.14) expresses the wind speed time history as the sum of many uncorrelated component functions of the form $e^{i\omega} = cos(\omega t) + i\,sin(\omega t)$, each associated with a small interval $d\omega$ on the frequency axis and each multiplied by a spectral amplitude $S_u(\omega) d\omega$. The frequency decomposition representation of the stochastic process provides a foundation for simulation of wind speed time histories or any other stochastic process of interest (Shinozuka and Jan 1972; Deodatis 1996; Shinozuka and Deodatis 1997; Senthooran et al.

2004; Carassale and Solari 2006).

The wind spectra information (Simiu and Scanlan 1996) is related to meteorological conditions and dependent on the geographical locations, the elevation levels, etc. The van der Hoven spectrum (van Der Hoven 1957) was the first comprehensive spectrum compiled that showed the components and characteristics of the wind in the atmospheric boundary layer in terms of frequency. In many wind engineering applications it is normal to examine only the micrometeorological conditions (higher frequency region of the van der Hoven spectrum: $10^{-3}$ to 1 Hz). There are many mathematical forms that have been used for wind spectra in meteorology and wind engineering. The most common and mathematically correct of these for the longitudinal velocity component (parallel to the mean wind direction) is the von Karman/Harris form (developed for laboratory turbulence by von Karman (1948), and adapted for wind engineering by Harris (1968)). The von Karman spectrum is commonly used in non-dimensional form as

$$\frac{f.S_u(f)}{u_*^2} = \frac{24\left(\frac{fL_u}{\bar{U}(10)}\right)}{\left[1+70.8\left(\frac{fL_u}{\bar{U}(10)}\right)^2\right]^{5/6}} \tag{2.15}$$

where $\bar{U}(10)$ = the mean wind speed at the height of $10$ m; $f$ = the frequency; $L_u$ = a turbulence length scale, which is a measure of the average size of the turbulent eddies of the air flow; $u_*$ = the friction velocity, which reflects the roughness of the ground surface and is related to the shear stress constitution in air flow at the surface. Based on results of a study of about 70 spectra of the horizontal components of gustiness in storing wind, Davenport (1961) suggested the following expression for the spectrum of horizontal gustiness as

$$\frac{f.S_u(f)}{u_*^2} = \frac{4\left(\frac{fL_u}{\bar{U}(10)}\right)^2}{\left[1+\left(\frac{fL_u}{\bar{U}(10)}\right)^2\right]^{4/3}} \tag{2.16}$$

in which $L_u$ =1200m. It is noted that both the von Karman and Davenport spectrum do not reflect the dependence of spectra on height. The height dependent wind spectrum was proposed by Kaimal (1972) as

$$\frac{f \cdot S_u(z,f)}{u_*^2} = \frac{200 \left( \dfrac{fz}{\bar{U}(z)} \right)}{\left[ 1 + 50 \dfrac{fz}{\bar{U}(z)} \right]^{5/3}} \qquad (2.17)$$

where $\bar{U}(z)$ = the mean wind speed at the height of $z$ m. The three sets of wind spectra are plotted in Figure 2.3. The spectrum comparison shows that the Davenport spectrum may overestimate the longitudinal spectra of turbulence in the higher frequency range of tall building serviceability check (e.g., $f > 0.1$ Hz) by as much as 100-400%, as shown in Figure 2.3. In the lower frequency range (e.g., $f < 0.01$ Hz), the Davenport spectrum tends to underestimate the spectral value. Since the spectral distribution in the lower frequency range has little influence on tall building response, the Davenport spectrum is still widely used in wind engineering and design wind codes.

Davenport (1961, 1963, 1964) firstly developed a spectral approach to the wind-induced vibration of structures based on random vibration theory. Other early contributions to the development of this spectral approach were made by Harris (1963) and Vickery (1966). The wind-tunnel based semi-analytical random vibration approach in the frequency domain has become a common practice to predict wind loads and responses of tall buildings (Davenport and Isyumov 1967; Reinhold 1982; Tschanz and Davenport 1983; Vickery and Daly 1984; Cermak 1977, 2003). By means of either the high-frequency force balance (HFFB) or synchronous multi-pressure sensing system (SMPSS), aerodynamic wind loads can be estimated experimentally on a rigid scale model of the prototype. Based on the measured aerodynamic wind load, the dynamic response of a building system can then be analyzed using spectral approach in the frequency domain. Based on wind tunnel measurements of wind loads, the dynamic analysis of wind-induced lateral-torsional motion of asymmetric buildings has been studied in the frequency domain by a number of researchers. Tallin and Ellingwood (1985) developed a method to relate alongwind, crosswind, and torsional forces acting on square isolated buildings to building accelerations. The effects on building motion of statistical correlations between components of wind forces and mechanical coupling of the modes due to eccentricities of mass and rigidity were examined and clarified. Based on the equations of motion derived for asymmetric coupled buildings, Kareem (1985) presented a random-vibration based procedure for estimating the wind-induced lateral-torsional response of tall buildings, and examined the effects due to eccentricities in centers of mass and stiffness on the coupled motion of the building. For structurally asymmetric buildings having closely spaced translational and torsional natural frequencies, the statistical correlation between the crosswind and torsional motions has been found to play a significant role in determining the wind-induced response (Islam et al. 1992).

Although the HFFB technique has been used widely for quantifying generalized wind forces on buildings, it is primarily suitable for buildings with 1D mode shape in each principal direction. The limitations inherent in the HFFB technique have been discussed and highlighted especially on its applicability to buildings with three-dimensional (3D) mode shapes (Yip and Flay 1995). To overcome these limitations, much effort has been made to refine force balance data analysis techniques to take into account the coupled motion of tall buildings with 3D mode shapes (e.g., Yip and Flay 1995; Holmes et al. 2003; Flay et al. 2003). Chen and Kareem (2005a) presented a new systematic framework for dynamic analyses of 3D coupled wind-induced responses of buildings and the determination of equivalent static wind loads (ESWL) by representing the building response into the mean, background and resonant components. Both the cross correlation of wind loads acting in different directions and the intermodal coupling of modal response components were taken into consideration in the analysis and modeling of ESWL. Utilizing a representative tall building with 3D mode shapes, Chen and Kareem (2005b) further examined the dynamic wind effects on asymmetric buildings with closely spaced natural frequencies and highlighted the significance of cross correlation of wind loads and the intermodal coupling of modal responses on the accurate prediction of coupled building responses.

Due to the stationarity of wind speed fluctuation, wind loads and responses are correspondingly stationary and are sufficiently described by PSD functions in the frequency domain. However, the flow field created by some extreme wind events, e.g., thunderstorm, varies significantly from the traditional ABL stationary and homogeneous wind flows (Letchford et al. 2001). Characterization and modeling of transient nonstationary winds and their effects on structures have been received increasing attention in recent years (Holmes and Oliver 2000; Letchford et al. 2001; Chen and Letchford 2004a and 2004b; Holmes et al. 2005). Based on the evolutionary PSD (EPSD), Chen (2007) presented a frequency domain analysis framework for quantifying alongwind tall building response to transient winds. In that study, the transient winds and associated wind loads in buildings are modeled as the sum of deterministic time varying mean and evolutionary random fluctuating components, as

$$U'(t) = \bar{U}(t) + u'(t) \qquad (2.18)$$

The evolutionary PSD of $u'(t)$ could be related to $S_u(\omega)$ as

$$S_{u'}(\omega,t) = |a(t,\omega)|^2 S_u(\omega) \qquad (2.19)$$

where $a(t,\omega)$ = a complex-valued deterministic modulation function of both $t$ and $\omega$. The extension of spectral representation to nonstationary random process is due to Priestley (1965). Given the EPSD information, the time history is also possible to be simulated using the computationally efficient cosine series formula as demonstrated by Liang et al. 2007. That paper presents a rigorous derivation of a previously known cosine series formula (Shinozuka and Jan 1972) for simulation of one-dimensional, univariate, nonstationary stochastic processes integrating Priestly's evolutionary spectral representation theory.

## 2.3   Structural design optimization

### 2.3.1   Classical optimization method

Since the 1960s, various methods and formulations for optimization of problems in many diverse fields, such as physics, structural and mechanical engineering, operation research, economics and finances and others, have been developed and discussed in the literature. Although optimization applications in each scientific discipline are quite different, the fundamentally mathematical concept is essentially the same. In mathematics, the term optimization, or mathematical programming (MP), refers to the study of problems in which one seeks to minimize or maximize a real function by systematically choosing the values of real or integer variables from within an allowed set. This problem can be represented as follows

Given $f(\mathbf{x}) = f(x_1, x_2, \cdots x_n)$   $\mathbf{x} \in \mathbf{R}^n$

Find $\mathbf{x}^*$

Such that $f(\mathbf{x}^*) \leq f(\mathbf{x})$   or   $f(\mathbf{x}^*) \geq f(\mathbf{x})$ for all $\mathbf{x}$

Problems formulated using this technique in the fields of physics may refer to the technique as total-energy minimization, speaking of the value of the function $f$ as representing the energy of the physical system being modeled (Payne et al. 1992). The iterative minimization technique for total-energy calculation developed by Payne et al. became one of the standard methods in Computational Molecular Dynamics (CMD).

For twice-differentiable functions, unconstrained problems can be solved by finding the points where the gradient of the objective function is zero (that is, the stationary points) and using the Hessian matrix to classify the type of each stationary point. If the Hessian is positive definite, the

27

function value at the point is a local minimum, if negative definite, a local maximum, and if indefinite it is some kind of saddle point, as shown in Figure 2.4. However, existence of derivatives is not always assumed and many methods were devised for specific situations.

The basic classes of methods, based on smoothness of the objective function, are: Zero-order (Derivative-free) methods; First-order methods; Second-order methods. The golden section and polynomial fitting techniques are commonly used as zero order methods for one-dimensional minimization. For $n$-dimensional minimization, the conjugate direction method (Powell 1964) is an efficient method for finding the minimum of a function of several variables without calculating derivatives. As a first-order method, the steepest descent method, which searches for the optimum along the gradient vector, is usually more efficient than zero-order method. Although the steepest descent method is elegant and simple, the convergence rate of the method for functions form an elongated design space is very poor. Fletcher and Reeves (1964) improved the steepest descent method and convert it to the conjugate gradient method, which takes account the trend information of the previous search direction while including the current gradient information. Newton's methods are second-order methods which require second-order derivative information. Newton-Raphson method requires the gradient vector and the Hessian matrix information, which may not necessary always be available. Quasi-Newton or Variable-metric method (Kirsch 1993) uses the approximate forms of the Hessian matrix and its inverse. Davidson-Fletcher-Powell (DFP) update procedure builds up approximate inverse of the Hessian matrix using only the first derivatives (Fletcher and Powell 1963). Broyden-Fletcher-Goldfarb-Shanno (BFGS) update procedure builds up approximate Hessian matrix rather than its inverse (Kirsch 1993).

Constrained problems can often be transformed into unconstrained problems with the help of Lagrange multipliers $\lambda$. For a general minimization problem of $f(\mathbf{x})$ subject to ($j=1,2,...,m$) inequality constraints $g_j(\mathbf{x}) \leq 0$ and ($k=1,2,...,l$) equality constraints $h_k(\mathbf{x}) = 0$, the unconstrained Lagrangian function could be defined as

$$L(\mathbf{x},\lambda) = f(\mathbf{x}) + \sum_{j=1}^{m} \lambda_j g_j(\mathbf{x}) + \sum_{k=1}^{l} \lambda_{m+k} h_k(\mathbf{x}) \qquad (2.20)$$

By differentiating the Lagrangian function with respect to each design variable $x_i$ setting the derivatives to zero, the Karush-Kuhn-Tucker (KKT) necessary conditions are obtained as (Luo et al. 1996)

$$\frac{\partial f(\mathbf{x})}{\partial x_i} + \sum_{j=1}^{m} \lambda_j \frac{\partial g_j(\mathbf{x})}{\partial x_i} + \sum_{k=1}^{l} \lambda_{m+k} \frac{\partial h_k(\mathbf{x})}{\partial x_i} = 0 \quad (i=1,2,...,n) \qquad (2.21)$$

The KKT conditions are generally a nonlinear system of equations, and may not be easy to solve the system analytically. In practice, the numerical analysis technique (e.g., the Gauss-Seidel iterative technique) is always employed to solve the nonlinear system of equations (Householder 2006).

### 2.3.2 Formulations for structural optimization

#### 2.3.2.1 Mathematical Programming approach

In the structural optimization literature, two basically different formulation approaches for optimum design have been developed since 1960s, named as Mathematical Programming (MP) approach and Optimality Criteria (OC) approach. Schmit (1960) was the first to offer a comprehensive idea of structural synthesis of integrating finite element structural analysis and MP techniques to automate optimal elastic design. MP approach is attractive due to its generality and rigorous theoretical basis. All classical optimization method discussed in previous section could be applied in the MP approach (Kirsch 1993). The OC approach is discussed in the next section.

Traditionally, the finite element structural analysis and design optimization are performed sequentially in the MP approach. That is to say, only the design variables are treated in the formulation of optimization, while the state variables of systems, such as displacements, accelerations, strains, stresses and internal forces, are evaluated by the finite element structural analysis and treated as implicit functions of the design variables. Therefore, the design sensitivity analysis (Arora 1995; Choi and Kim 2005, Huang and Chan 2007) is needed to perform before optimization for obtaining the sensitivity information of state variables with respect to design variables, which will be demanded by the optimization algorithms. The finite difference methods have been used to calculate the gradients since they are easy to implement using FEM program. However, the finite difference methods have accuracy problems, i.e., the so-called "step-size" dilemma (Haftka and Gurdal 1992). Given the global stiffness matrix of a structural system, the direct differentiation method or the adjoint variable method can be employed to analytically determine the design sensitivity information (Choi and Kim 2005). Based on the principle of virtual work, Huang and Chan (2007) presented a design sensitivity analysis method for the displacement and eigenvalue sensitivities without the need for the global stiffness matrix information of a building system. It is worth to note that the latest development on efficient structural approximation and reanalysis methods (Kirsch 2000, 2003) could directly be used to quantify various desired sensitivity information within the optimization framework.

When the state variables are also treated in the formulation of optimization by formulating the system governing equilibrium equations as equality constraints, the MP approach becomes a general class of formulations known as mathematical programming with equilibrium constraints (MPECs) with simultaneous analysis and design for the optimization problems (Luo et al. 1996; Arora and Wang 2005). The equilibrium constraints could be expressed in terms of functional based on the variational principle in physics to model the equilibrium phenomenon in physics, mechanics, engineering and other applications. Although the MPEC provide a general approach to formulate various optimization problems in wide disciplines, the MPEC formulation may be posed as a non-convex and non-differentiable optimization problem, which is computationally difficult to solve (Luo et al. 1996; Ferris and Pang 1997). Optimization algorithms have to be explored and devised for different problems posed by MPECs (Hilding et al. 1999; Ferris and Tin-Loi 2001; Evgrafov and Patriksson 2003).

### 2.3.2.2 Optimality Criteria approach

As an alternative approach, the Optimality Criteria (OC) approach were emerged and developed in the late 1960s and early 1970s. Prager et al. (1968) presented the OC method for optimal structure design in analytical forms, while Venkayya et al. (1968) further developed the OC approach using numerical analysis technique for practical large scale aerospace structures. Based on the way to obtain the optimality criteria, the OC method can be classified into the intuitive OC method and the rigorously derived OC method.

An early example of an intuitive OC approach is the Fully Stressed Design (FSD) method, which is applicable to structures subject only to stress and sizing constraints. Although the FSD method is effective as a practical design tool, the FSD method only can achieve true optimum for statically determinate structures but not for statically indeterminate structures. For these indeterminate structures, the minimum weight design may not always be fully stressed (Schmit 1960; Razani, 1965; Panagiotis et al. 2002). Panagiotis et al. 2002 somewhat overcame the drawback of the FSD method by introducing a strain energy criterion. Pedersen et al. (2005) develop an optimality criterion in eigenvalue problems for shape optimization of plate. The work extends the application area of OC method from the sizing optimization to shape optimization. The rigorously derived OC method was introduced by Cheng and Truman (1983), Chan (1992), Chan et al. (1995) to civil engineering by solving the optimal design problems of building structures. The rigorously derived OC method mainly consists of two complementary phases. The first phase concerns the derivation of a set of necessary conditions (KKT conditions) that the optimal design must satisfy. The second phase involves the application of a numerical recursive algorithm to vary design variables in order to satisfy the KKT conditions for the final optimum design solution. Unlike the MP method, the OC method is more computationally efficient, and usually provided a local optimum design with a few analysis-and-design cycles.

Chan (1992) developed a virtual work formulation to implement the OC method for the design optimization of tall steel buildings. In that study, the OC method has been formulated based on the principle of virtual work, by which the displacement of a tall building could be fairly approximated. Since a tall building, in a global sense, is a statically determinate cantilever structure, internal element force distributions in statically indeterminate tall building structures are relatively quite insensitive to moderate changes in the element sizing variables. Therefore, the virtual work formulation represents a good approximation of the behavior of the drift constraints and the explicitly expressed drift formulation generally leads to smooth and rapid solution convergence. Another advantage of the virtual work formulation OC approach is easy to apply for the optimum design of practical tall building structures, which are generally modeled by the commercial FEM software. The information of internal element forces demanded by the virtual work formulation is readily available from the FEM model of a tall building structure.

Much research effort has been continually devoted to the OC method and its application due to its high efficiency and quick convergence behavior (Chan 1997, 1998, 2001; Chan and Zou 2004; Chan and Chui 2006; Chan et al. 2007). Based on the virtual work formulation (Chan 1992; Huang and Chan 2007), the OC method was employed to solve the optimization problem of large scale tall buildings subject to multiple equivalent static wind drift and wind-induced vibration constraints (Chan 2001). Latter, the OC method was extended to solve the optimal performance-based seismic design problems of reinforced concrete buildings (Chan and Zou 2004; Zou and Chan 2005). Chan et al. (2007) presented an integration of an aerodynamic wind load analysis and an optimal element resizing technique based on the OC method for lateral drift design of tall steel buildings.

While quick convergence can be normally achieved using the OC technique, it cannot always assure that the global optimum can be found. Advances in recent research have resulted in a hybrid methodology, namely the OC-GA method, which incorporates Genetic Algorithms (GAs) into the gradient based OC technique. The evolutionary GAs are in general more robust and present a better global behaviour than the OC. However, GAs alone may suffer from a slow rate of convergence towards the global optimum. In order to benefit from the advantages of both OC and GAs, a hybrid combination of both methodologies has been developed and thus the so-called OCGA method as an

31

attempt to improve the robustness as well as the computational efficiency of the optimization procedure for structural topology and element sizing design of tall buildings (Chan and Wong 2007).

## 2.3.3 Dynamic response optimization

### 2.3.3.1 Optimization of structures under transient loads

In this study, the optimization of structures subject to transient loads in a time domain is referred as dynamic response optimization (Kang et al. 2006). Unlike optimizations of structures under static loads, the dynamic response optimization would involve numerous time-dependent constraints. The treatment of time-dependent constraints becomes one of key topics in the dynamic response optimization. The other two important topics are sensitivity analysis and approximation technique. While the time-dependent constraint treatment is unique for the dynamic response optimization, the sensitivity analysis and approximation are the equally important issue for optimization of structures under static loads, as discussed in previous section.

Much research has been focused on the treatment of numerous time-dependent constraints imposed in the dynamic response optimization problems (Hsieh and Arora 1984, 1985, 1986; Grandhi et al. 1986). Haug and Arora (1979) converted numerous time-dependent constraints into one single equivalent functional constraint by integrating original point-wise constraints over the entire time interval of interest. Hsieh and Arora (1984) presented the worst-case design formulation as the constraint is imposed on the worst response. In this treatment of time-dependent constraints, all point-wise time-dependent constraint are replaced by several constraints that are imposed at the multiple local maximum response points, or just simply by one single constraint that is imposed at the global maximum response point. A hybrid approach, so call sub-domain functional formulation, for treating time-dependent constraints has been discussed and evaluated by Tseng and Arora (1999). The hybrid treatment of point-wise performance constraints for the dynamic response optimization problem is to transform point-wise constraints into several equivalent functional constraints by dividing the entire time domain into several sub-domains. Using the worst-case design formulation, Zou and Chan (2005) developed an effective numerical technique for member sizing optimization of concrete building frameworks subject to seismic drift design constraints under response spectrum and time history loadings.

The direct differentiation method and the adjoint variable method as discussed for static structural design optimization are still applicable to the sensitivity analysis for the dynamic response optimization problem. However, in dynamic response optimization, the system behavior is no longer only related to stiffness, mass and damping would also play important roles in the governing equations of motion. The direct differentiation method has been applied to equations of motion and solved by numerical time-stepping methods (Newmark 1959; Greene Haftka 1991; Sousa et al. 1997). As an efficient dynamic analysis method, the mode superposition method has also been employed in sensitivity calculations in linear transient structural analysis (Choi et al. 1983; Wang 1991; Lee 1999). The adjoint variable method is introduced by defining an augmented response function to calculate sensitivity of a dynamic response (Arora and Cardoso 1989; Lee 1999; Kocer and Arora 2002).

Since the dynamic analysis in time domain is expensive in large scale structural optimization, much effort to approximate the dynamic response of structures has been made. Approximation can be classified into three categories as global approximation, local approximation and combined approximation (CA). Response surface method (RSM) is the representative one of global approximation (Montgomery 2001; Kim and Choi 2007). RSM builds an approximated response model from data acquired at various design points by using statistical approaches. The built approximation model is usually an explicit function of the design variables and is easy to handle sensitivity analysis (Kurtaran et al. 2002). Unlike the global approximations obtained by analyzing the structure at a number of design points, local approximations are based on information calculated at a single design point. Jensen and Sepulveda (1998) applied the local approximation by introducing intermediate variables to calculate design sensitivity of transient response of structures.

The combined approximation (CA) method, which attempts to give global qualities to local approximations, has been developed recently and applied for nonlinear and dynamic reanalysis, as well as dynamic sensitivities (Kirsch 2003; Kirsch et al. 2007; Kirsch and Bogomolni 2007). The CA approach is based on the integration of several concepts and methods in numerical analysis, including series expansion, reduced basis, matrix factorization, and Gram-Schmidt orthogonalizations. The advantage is that efficient local approximations and accurate global approximations are combined to achieve an effective solution procedure. Bogomolni et al. (2006) demonstrated that dynamic sensitivities can be efficiently calculated using the CA method and finite difference methods for the discrete linear systems subjected to dynamic loading.

The dynamic response optimization discussed so far solves differential equations to calculate design sensitivity, which is relatively computationally expensive compared to the static response sensitivity calculation. In order to exploit well-established static response optimization techniques in dynamic response, Kang et al. (2001) introduced the equivalent static load (ESL) method in dynamic

33

response optimization by transforming dynamic loads into static loads based on displacement equivalence. Choi and Park (2002) developed a quasi-static optimization method for the structure under dynamic loads. The sets of ESL are generated at all time intervals and utilized as a multiple loading condition in the optimization process. Actually, the transformation of dynamic loads into equivalent static loads (ESLs) has been well developed in civil engineering for analysis and design of building and bridge structures under earthquake or typhoon loads (Chopra 1995; Davenport 1967, 1995; Holmes 2002; Chen and Kareem 2005a; Chan et al. 2007). Due to its simplicity and efficiency, current design codes adopt the ESLs approach to treat the intrinsic random and dynamic loads, e.g., the response spectrum analysis method for earthquake loads, the gust loading factor method for wind loads. The dynamic optimal seismic or wind-resistant designs using ESL approach or any other are discussed in next two sections, respectively.

### 2.3.3.2 Seismic structural optimization

Much research on the optimal seismic design has been developed even before the establishment of performance-based design concepts, i.e., multiple performance levels and design objectives for buildings under earthquake loading. Bhatti and Pister (1981) considered an optimization problem for the earthquake-resistant design of structural systems that incorporated two levels of design constraints: The first level is for frequent minor earthquake for which the structure are remained in the linear range; the second design level is for less frequent, major earthquakes, for which inelastic deformation and limited damage are allowed. Pantelides (1990) studied the optimum design of active control seismic structures by optimizing the structural members of frames equipped with an active control system. Optimization of reinforcement concrete (RC) buildings against earthquake loads has been presented by Truman and Cheng (1997).

With increasing popularity of performance-based design approach, many researchers and engineers have proposed various methodologies, which are aimed to incorporate performance-based concepts and criteria into the optimization framework. Foley (2002) presented an overall literature review on optimal performance-based deign approach. Ganzerli et al. (2000) addressed the optimal performance-based design of seismic structures. In that study, performance-based design concepts and pushover analysis were incorporated into the optimization method for reinforced concrete structures. Chan CM and Zou XK (2004) presented an effective optimization technique for the elastic and inelastic drift performance design of reinforced concrete buildings under response spectrum loading and pushover loading. The equivalent static load (ESL) method has been used for seismic performance-based design optimization (Gong et al. 2005; Xu et al. 2006). A performance-based design sensitivity analysis procedure for inelastic steel moment frameworks

34

under equivalent static earthquake loading (ESEL) was developed by Gong et al. 2005. Such a sensitivity analysis procedure was later integrated into an optimal performance-based seismic design framework by Xu et al. 2006.

### 2.3.3.3  Wind-resistant structural optimization

For multistory buildings, the equivalent static wind loads (ESWLs) corresponding to the specific incident wind angle are generally expressed in terms of the alongwind, crosswind and torsional directions; and each directional ESWL consists of the mean, background and resonant components. For normal low-rise buildings, i.e., less than 100m, wind-induced structural loads are dominated by static mean components and quasi-static background components. The wind-induced resonant effects on low-rise buildings are small and negligible such that wind loads can be considered as constant static design loads. The well developed static response optimization technique can be directly employed in wind-resistant structural optimization (Chan 1992, 1997, 1998). However, for dynamically sensitive tall buildings, wind-induced resonant effects become critical. In order to make an accurate prediction of the wind-induced structural loads on the building, it is necessary that the ESWLs be always updated whenever there exists a significant change in the structural properties of the building.

Recently, a promising and effective integrated wind-induced response analysis and serviceability optimization approach is developed to achieve an optimal design solution of wind-sensitive complex tall buildings (Chan and Chui 2006; Chan et al. 2007). Chan and Chui (2006) presented an occupant comfort wind-induced acceleration design optimization technique for wind-excited symmetric tall steel buildings based on eigenvalue sensitivity analysis using Rayleigh method. Chan et al. (2007) developed an integrated optimal design framework that couples together an aerodynamic wind load updating analysis procedure and a stiffness optimization method for symmetric tall building structures subject to serviceability drift design constraints. When given with the aerodynamic wind load spectra, ESWLs on tall buildings can be updated instantaneously for any change in the member sizes of a structural system during the optimization process (Chan et al. 2007). Although these studies represent a major advance in the use of structural optimization techniques for wind-induced serviceability design of tall buildings, it is necessary to extend the stiffness design optimization technique for general asymmetric tall buildings subject to multiple wind induced drift and acceleration constraints together.

## 2.4  Reliability-based design optimization

## 2.4.1  Reliability analysis method

The structural systems consisting of interacting and functionally interconnected elements are designed in order to function or operate in a normal condition without failure or in an extreme condition with a small probability of failure. In reliability theory, failure is any event, which is defined as the violation of a prescribed limit state by the state of an object (a system or an element). The limit state concept leads to limit state design philosophy, which has been adopted in the current machine or building design codes. In reality, the vast uncertainties always exist in systems or elements and their external environment or loadings. The basic uncertain quantities could be described as time-invariant random variables. Consequently the failure probability is time independent and the probability of no failure is time-invariant reliability. If the basic uncertain quantities are time-dependent and modeled as random processes, the failure probability is also a function of time. The time-variant probability of failure is defined as the first out-crossing of the state process through the limit state during a given time interval. The probability of failure-free operation of an object during the given time interval could be called as time-variant reliability.

### 2.4.1.1  Time-invariant reliability

The time-invariant probability of failure could be defined as

$$P_f = \int_{g(\mathbf{x}) \leq 0} f(\mathbf{x}) d\mathbf{x} \quad \mathbf{x} \in \mathbf{R}^n \tag{2.22}$$

where $\mathbf{x}$ is a vector of random variables that represents uncertainty quantities of systems and loadings; $f(\mathbf{x})$ is the joint probability density function of the random vector $\mathbf{x}$; $g(\mathbf{x}) \leq 0$ defines the failure domain $D_f$. A great deal of effort in the past several decades has been devoted to developing efficient algorithms for computing the multidimensional integrals of the form in Eq. (2.22) analytically or numerically (Der Kiureghian 1996; Rackwitz 2001).

The first-order reliability method (FORM) is the earliest and important approximate approach to compute reliability (Shinozuka 1983). In order to takes advantage of the rotational symmetry of the standard normal space, the FORM analysis require a transformation of limit state function $g(\mathbf{x})$ from the original parameter space into the new space of the uncorrelated standard normal variates, as $G(\mathbf{u}) = g[\mathbf{x}(\mathbf{u})]$. The transformation task is usually performed by using the so-called Rosenblatt-transformation (Rosenblatt 1952), which is based on the representation of a multidimensional distribution function by a product of conditional distributions. The simple result of the FORM method then is given by

$$P_f \approx \Phi(-\beta) \qquad (2.23)$$

where $\Phi(\cdot) =$ the Gaussian cumulative distribution function; $\beta = \|\mathbf{u}^*\|$, known as the reliability index, is the distance from the origin to the most probable failure point (MPFP) or "design point", which is a point located on the limit state surface with minimum distance form the origin, has the highest probability density among all failure points in the standard normal space. This point $\mathbf{u}^*$ is the solution of the following constrained minimization problem as

Minimize: $\|u\| = \mathbf{u}^T \mathbf{u}$

Subject to: $G(\mathbf{u}) = 0$

Then, the main computational task of the FORM is the location of "design point" by an optimization method (Liu and Der kiureghian 1990). One suitable method is the Lagrange multiplier method by defining a unconstrained Lagrangian function of the form in Eq. (2.20).

In FORM, the limit state surface is replaced by the tangent hyperplane at $\mathbf{u}^*$. It is natural to attempt a second-order expansion of the limit state surface based on the assumption that a second-order expansions of the limit state surface is better than a first-order expansion. Based on asymptotic analysis Breitung (1984) developed a second-order reliability method (SORM) to approximate the failure probability in terms of $\beta$ and the principal curvatures $\kappa_i$, $i = 1, ..., n - 1$ of the parabolic failure surface at u* as

$$P_f \approx \Phi(-\beta) \prod_{i=1}^{n-1} (1 - \beta\kappa_i)^{-1/2} \qquad (2.24)$$

Der Kiureghian et al. (1987) suggest a more computationally efficient "point-fitted" SORM procedure compared to the original "curvature-fitted" SORM. The exact result for the probability content of the failure domain obtained from the full second-order Taylor expansion of the failure function at the "design point" was presented by Tvedt (1990). In stead of working in the standard normal space, reliability approximation was also achieved by maximizing the log likelihood $l(\mathbf{x}) = \ln f(\mathbf{x})$ in the original variable space (Breitung 1991; Papadimitriou et al. 1996). The log likelihood maximization approach is built on the asymptotic analysis to Laplace-type integrals (Bleistein and Handelsman 1986).

## 2.4.1.2 Time-variant reliability

If the performance state of an object is a random process $\mathbf{Z}(t)$, the time-variant failure event is defined as the first out-crossing of $\mathbf{Z}(t)$ through the limit state surface $g(\mathbf{x}, \mathbf{Z}) = 0$. Denoting the number of excursions during time interval $[0,t]$ as $N(t)$, the time-variant probability of failure-free operation could be written in terms of the conditioned probability for given $\mathbf{x}$ based on the total probability theorem as

$$P_s(t) = \int P\{N(t) = 0 | \mathbf{x}\} f(\mathbf{x}) d\mathbf{x} \quad \mathbf{x} \in \mathbf{R}^n \tag{2.25}$$

where $P\{N(t) = 0 | \mathbf{x}\}$ is the conditional probability of no out-crossing during the time interval $[0,t]$.

On the other hand, the failure free operation means that within the time interval the state of the object are always within the safe domain $D_s$ defined by $g(\mathbf{x}, \mathbf{Z}) > 0$. Therefore, the time-variant reliability could be rewritten as

$$P_s(t) = \int P\{\mathbf{Z}(t) \in D_s | \mathbf{x}\} f(\mathbf{x}) d\mathbf{x} \quad \mathbf{x} \in \mathbf{R}^n \tag{2.26}$$

where $P\{\mathbf{Z}(t) \in D_s | \mathbf{x}\}$ is a conditional probability of failure free operation during the time interval $[0,t]$.

One of most appropriate model for the description of failure events of highly reliable systems is the Poisson distribution, i.e., $P_k(t) = \dfrac{(vt)^k}{k!} \exp(-vt)$. Poisson distribution is a discrete probability distribution that expresses the probability of $k$ failure events occurring in a fixed period of time $t$ if these events occur with a known average rate $v$ and independently of the time since the last event. Therefore, the conditional probability of failure-free operation is given by letting $k=0$ in the expression of Poisson distribution as

$$P\{N(t) = 0 | \mathbf{x}\} = \exp(-vt) \tag{2.27}$$

where $v =$ mean failure rate, which means that the expected number of failure events per unit time. The occurrence of failure events means the random state process $\mathbf{Z}(t)$ is crossing the limit state, i.e., simply defined by double barrier of level $b$ in the context of the classical first-passage problem (Vanmarcke 1975). The first result of mean failure (out-crossing) rate for a random scalar process was given by Rice (1945) as

$$v_b = \int_{-\infty}^{\infty} |\dot{z}| f_{Z,\dot{Z}}(b, \dot{z}) d\dot{z} \tag{2.28}$$

38

where $f_{Z,\dot{Z}}(b,\dot{z})$ is the joint probability density function of the scalar process $Z(t)$ and its derivative process $\dot{Z}(t)$.

For a nonlinear limit-state surface, a generalization of the Rice formula for a vector stochastic process is available that requires integration over the surface (Belyaev 1968). Based on the generalized Rice formula, the joint first-passage probability of a vector random process can be estimated (Song and Der Kiureghian 2006). The level crossings of a one-dimensional stochastic process has also been generalized to an $n$-dimensional filed $Z(\mathbf{t}), \mathbf{t} \in \mathbf{R}^n$ (Robert and Hasofer, 1976; Piterbarg, 1995). The high-level crossing problem naturally leads to the asymptotic extremal distributions for random process or field (Piterbarg, 1995; Muscolino and Palmeri 2005). The theorem on the asymptotic extremal distributions for homogeneous Gaussian fields has been applied to study the probability distribution of surface graivty waves in the open deep ocean (Socquet-Juglard et al. 2005). The relationship between the level crossing rate and the extreme value distribution for a random process is studied in detail in the context of wind-induced peak response of a tall building in Chapter 6.

A Markov process is a stochastic process in which the probability distribution of the current state is conditionally independent of the path of past states. Let the evolution of the state process $\mathbf{Z}(t)$ be a Markov process. Then the conditional probability of failure free operation could be rewritten in terms of transition probability density $f_{\mathbf{Z}}(\mathbf{z},t|\mathbf{Z}(t_0)=\mathbf{z}_0,\mathbf{x})$ for the realization of process $\mathbf{Z}(t)$ encompassed by the limit state surface as

$$P\{\mathbf{Z}(t) \in D_s|\mathbf{x}\} = \int_{D_s} f_{\mathbf{Z}}(\mathbf{z},t|\mathbf{Z}(t_0)=\mathbf{z}_0,\mathbf{x})d\mathbf{z} \qquad (2.29)$$

The transition probability density $f_{\mathbf{Z}}$ is governed by the FPK equation as (Lin and Cai 1995)

$$\frac{\partial f_{\mathbf{Z}}}{\partial t} = -\sum_{j=1}^{n}\frac{\partial}{\partial z_j}(a_j f_{\mathbf{Z}}) + \frac{1}{2}\sum_{j=1}^{n}\sum_{k=1}^{n}\frac{\partial^2}{\partial z_j \partial z_k}(b_{jk} f_{\mathbf{Z}}) \qquad (2.30)$$

where $a_j$ = the drift coefficient; $b_{jk}$ = the diffusion coefficient. These coefficients are determined from the equations of motion for a structural dynamic system. Very recently, the FPK equation method was successfully applied to obtain the probability density function of the nonlinear response of an articulated leg platform by Kumar and Datta (2008).

### 2.4.1.3 Simulation method

The Monte Carlo simulation method (Fishman 1996) provides approximate solution to a variety of mathematically formulated problem by performing statistical sampling experiments on a computer. The method, remarkably, applies to problems with absolutely no probabilistic content and to those with inherent probabilistic structure, e.g., reliability analysis. Although the standard Monte Carlo simulation (MCS) method offers a very robust solution to reliability problem, the computational requirement by MCS is quite prohibitively high when evaluating small failure probabilities (Proppe et al. 2003). Therefore importance sampling method was developed as an efficient simulation technique by sampling more frequently from inside the failure set (Harbitz 1986). The first-passage failure probability (time-variant reliability problem) of linear systems subjected to Gaussian white noise excitation was effectively estimated using importance sampling method (Au and Beck 2001). In that paper, a suitable importance sampling density was developed by studying the elementary failure region and elementary failure probability based on the impulse response function.

Other class of efficient simulation methods is based on the geometry properties of limits state functions and failure domains in the standard normal space. Der Kiureghian (2000) offered a new outlook to solve reliability problem for linear systems subject to Gaussian excitation by investigating the geometry of random vibrations. Katafygiotis and Cheung (2004) developed a wedge simulation method to calculating the reliability of linear dynamic systems through studying the interaction between the different linear elementary failure domains. Similarly, the domain decomposition method based on exploiting the special structure of the failure domain was proposed as a more efficient approach for estimating the first-passage probabilities of linear systems by Katafygiotis and Cheung (2006).

### 2.4.2   Reliability index optimization approach

In engineering practice, reliability analysis of mechanical and building systems is served as a key procedure in the design optimization of those systems with reliability constraints. Today, it is widely recognized that the design optimization method as a more rational and effective design tool should also account for the stochastic nature of engineering systems (Frangopol and Maute 2003; Schueller 2007). Frangopol and Maute (2003) presented a literature review for reliability-based design optimization (RBDO) of civil and aerospace structures. In that paper, reliability index approach is formulated as

Minimize:   $f(\mathbf{d})$

Subject to:   $\beta_j(\mathbf{x}, \mathbf{d}) - \beta_j^L \geq 0 \quad j = 1, ..., n_\beta$

where $\mathbf{x}$ = random variables, which represent various uncertain quantities rising from systems and loadings; $\mathbf{d}$ = design variables, which may be deterministic design parameters or statistical property values (e.g., mean and standard deviation) of random variables $\mathbf{x}$; $\beta_j^L$ = the lower limits of reliability index for the associated limit states prescribed. The relationship between the limiting value of reliability index and the target failure probability could be given by FORM as $P_f^T \approx \Phi\left(-\beta_j^L\right)$.

One of the first attempts for the development of RBDO was made by Moses in 1969. It was stated in his study which focused on the relationship between reliability and optimization, that "An optimization procedure which uses overall structural failure probability as the behavior constraint should produce more balanced designs, consistent with the development of rational safety." Since 1960s, much work has been done towards the goal sated by Moses (Surahman and Rojiani 1983; Frangopol 1985a, 1985b; Kim and Wen 1990; Enevoldsen et al. 1994; Chang et al. 1994; Chandu and Grandhi, 1995; Zou and Mahadevan 2006). Some of those literature, the reliability index approach was employed. Given the target failure probability, a reliability based optimization procedure for reinforced concrete structures was developed by Surahman and Rojiani (1983). The component or system reliability index based optimization problem is formulated by Enevoldsen et al. 1994. Then the problem is solved using the FORM in estimating the reliability and sequential quadratic programming techniques. Chang et al (1994) used the reliability index approach for optimum seismic design of steel structures subjected to seismic loadings of the Uniform Building Code. Zou and Mahadevan (2006) developed an efficient decoupling reliability-based design optimization method to solve a practical vehicle side impact design problem subject to multiple reliability index constraints.

### 2.4.3 Performance measure optimization approach

The reliability index constraint can be also formulated in terms of probabilistic performance measure constraint (Tu et al. 1999). The time-invariant failure probability given in Eq. (2.22) would be constrained during RBDO so as to be within the limit of target failure probability as

$$P_f = \int_{g(\mathbf{x},\mathbf{d})\leq 0} f(\mathbf{x})d\mathbf{x} \leq P_f^T \approx \Phi\left(-\beta_j^L\right) \qquad (2.31)$$

Since the statistical description of limit state function $g_j(\mathbf{x},\mathbf{d})$ is characterized by its cumulative distribution function (CDF) $F_{g_j}(t) = P\left(g_j \leq t\right)$, the probabilistic constraint of Eq. (2.31) can be rewritten as

$$0 \le F_{g_j}^{-1}\left[\Phi\left(-\beta_j^L\right)\right] = t^* \tag{2.32}$$

where $t^*$ = the performance measure. Therefore, Eq. (2.32) represents a probabilistic performance measure constraint. On the other hand, the performance measure constraint can be easily transformed back into the reliability index constraint as

$$\beta_j^* = -\Phi^{-1}\left[F_{g_j}(0)\right] \ge \beta_j^L \tag{2.33}$$

Lee et al. (2002) compared the target-performance-based approach with the conventional reliability-index-based approach for solving the structural design optimization problem. The illustrative examples in that paper demonstrated that the performance measure approach is superior to the reliability index approach in view of both computational efficiency and numerical stability. Youn et al. (2003) developed a hybrid analysis method to take advantages of the performance measure approach for effective evaluation of probabilistic constraints in the RBDO process. The performance measure concept is found to be effective to establish the CDF of output performance function for practical engineering applications, i.e., side-impact crashworthiness, involved numerous input uncertainties (Youn et al. 2005). It is worth to note that in the RBDO framework, depending on the characteristic of design problems, the trade-off between the accuracy and efficiency must be considered in selecting the appropriate reliability methods among various approximation or simulation approaches discussed in previous sections (Jensen 2005).

## 2.5 Summary

Based on literature review, a number of conclusions were drawn in the following:

1. The performance-based design approach seems to be a promising direction for designing economical and reliable building structures satisfying intended performance design objectives under high-risk hazard-induced loadings, e.g., wind and seismic loads. The source of typhoon risk and the method to predict it have been reviewed. Major uncertainties involved in wind-sensitive tall building design are identified, and the methods to model and quantify uncertainty are discussed.

2. Literature review revealed that the dynamic theory of linear systems is so well developed compared to that of nonlinear systems. Although many literature are available addressing the nonlinear dynamic analysis problems, there is still lacking a general and efficient approach to predict nonlinear response of a randomly excited stochastic system. Some approximation

methods with additional assumptions have to be adopted to solve practical engineering analysis and design problems. Wind-tunnel based semi-analytical method in frequency domain becomes a practical and effective way to predict wind-induced dynamic responses of tall buildings for the design purpose.

3. The OC approach has gained tremendous popularity for optimal building structural design due to higher efficiency than other classical optimization method or mathematical programming approach. The OC algorithm based on the virtual work formulation have been employed in this study as a key optimization engine in the computer based framework of reliability performance-based design optimization of wind-sensitive tall buildings.

4. Some methods for dynamic response optimization have so far been developed and studied. However, these methods are only feasible for mechanical structures subjected to deterministic transient time-history loads. Dynamic response optimization concepts have to be extended to deal with situations of random vibration in the optimal design of wind-sensitive tall buildings.

5. Reliability theory and reliability analysis methods have been extensively reviewed. Due to the complexity of multidimensional integral, most of reliability analysis problems have to be approximately solved using numerical analysis technique, including optimization algorithms and simulation. It is worth to note that the Markov model (consequently, FPK equations) is equally important to study nonlinear responses of randomly excited systems and time-variant reliability problems. For wind-induced time-variant reliability problem, the classical first-passage problem solution could offer a convenient way to quantify the serviceability failure probability. The latest development of efficient simulation methods for randomly excited linear system could also help to solve this reliability problem.

6. Much research effort has been devoted to the area of reliability-based design optimization aimed to making the design optimization method as a more rational and effective design tool by accounting for the stochastic nature of engineering systems. However, a practical design method for optimal stiffness design of wind-sensitive tall buildings subject to deterministic and probabilistic dynamic serviceability constraints is still lacking.

Throughout the process of literature review, it is believed that the fulfillment of the objectives presented in Section 1.2 could be achieved by investigating some of the key research components regarding the wind-induced dynamical analysis technique, dynamic response optimization, wind-induced time-variant reliability analysis and the reliability performance-based optimization framework. This comprehensive study would have provided practical and useful design tools for structural and wind engineers to design an economical and reliable building structure, which is safe over its intended life and serviceable for its intended function.

43

**Table 2.1   Recommended performance-based wind engineering design level**

| Wind design severity level | Average return period | Probability of exceedance | Performance levels |
|---|---|---|---|
| Very Frequent | 1 year | 100% in 50 years | Perception threshold |
| Frequent | 5 years | 99.9% in 50 years | Occupant comfort |
| Frequent | 10 years | 99.5% in 50 years | Fear for safety |
| Occasional | 50 years | 64% in 50 years | Drift / Strength |
| Rare | 475 years | 10% in 50 years | Safety |
| Very Rare | 1000 years | 5% in 50 years | Collapse prevention |

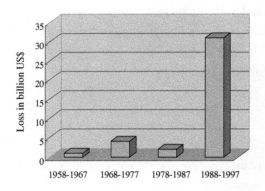

**Figure 2.1   National earthquake related losses in United States**

(Excerpt: FEMA 349)

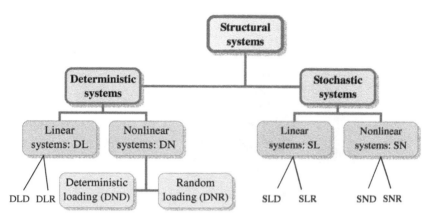

**Figure 2.2   Classification of the dynamic analysis problems of structural systems**

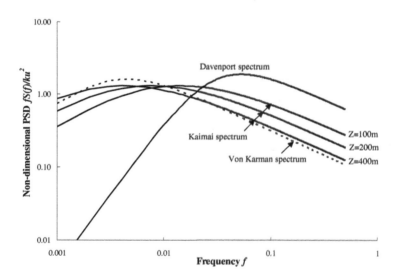

**Figure 2.3   Spectrum of horizontal gustiness**

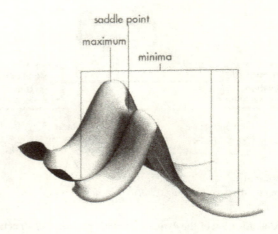

**Figure 2.4    Stationary points of a smooth function**

# CHAPTER 3　Coupled Dynamic Analysis of Wind-excited Tall Buildings

## 3.1　Introduction

Modern tall buildings are wind sensitive structures. Recent trends towards developing increasingly taller and irregularly-shaped complex buildings imply that these structures are potentially more responsive to wind excitation. Making accurate predictions of wind loads and their effects on such structures is therefore a necessary step in the design process. Contemporary tall buildings having complex geometric shapes accompanied by non-coincident centers of mass and resistance may vibrate in a lateral-torsional manner, thus making the prediction of wind-induced responses of such buildings more difficult than that of regular buildings with simple geometric shapes and unidirectional mode shapes (Tallin and Ellingwood 1985; Kareem 1985; Yip and Flay 1995; Chen and Kareem 2005a). As discussed in Chapter 2, the wind-tunnel based semi-analytical random vibration approach in frequency domain has become a common practice to predict wind loads and response of tall buildings. In the earlier work of Tallin and Ellingwood (1985) and Kareem (1985), the crosswind force and torque were assumed to be in phase and closely correlated such that the cross power spectra density (XPSD) of the crosswind and torsional loads was replaced by co-spectra (real part of XPSD) in the analysis. The implication of such an assumption on the dynamic response prediction was later investigated by Islam et al. (1992). It was found that the effect of the quad-spectra (imaginary part of XPSD) reflected by the phase angle of XPSD could not be neglected in the dynamic analysis in order to accurately predict wind-induced response of buildings with three-dimensional (3D) mode shapes.

Based on random vibration theory, the total dynamic response of a wind-excited building can be accurately evaluated using the full-spectral approach in the frequency domain, which fully takes into account the effects of all cross modal correlations as well as the mechanical coupling of 3D mode shapes. For the sake of simplicity, the full-spectral formulation can be reduced into a form of modal combination, such as the square root of sum of square (SRSS) combination rule and the complete quadratic combination (CQC) rule (Wilson et al., 1981). The traditional CQC factor (Der Kiureghian 1980) in the closed-form expression in terms of modal frequencies and modal damping ratios was derived based on the assumption of a single source of white-noise seismic excitation. However, the direct use of the traditional CQC factors for predicting wind-induced response of modern tall buildings may not necessarily be adequate (Xie et al. 2003; Chen and Kareem 2005b). Modern tall buildings with complex geometric shapes and asymmetric distributions of mass and stiffness may result in closely spaced fundamental frequencies, the strong dependence of modal response components found in these buildings do have a significant impact on the dynamic response of tall buildings (Der Kiureghian 1981), thus requiring a more accurate prediction of the intermodal correlation effects to facilitate the utilization of the CQC scheme in the framework of modal analysis for predicting wind-induced response. Research has also indicated that when either of the two modes has a relatively high frequency mode, the white-noise assumption may result in an underestimation of the modal correlations in structures under a single source of seismic excitation (Der Kiureghian and Nakamura, 1993). In addition, under multi-point wind excitation, the imaginary part of the XPSD (quad-spectra) among generalized forces may become significant in determining the wind-induced loads and the coupled response of tall buildings. Hence, it is necessary to investigate the validity of the white noise assumption and to study the impact of the imaginary part of XPSD among generalized forces on determining the cross correlations of modal response and subsequently the wind-induced response of tall buildings with 3D modes and closely spaced natural frequencies.

In this study, an accurate dynamic analysis framework capturing rigorously the cross correlations of modal responses of tall asymmetric buildings in wind-induced motions is presented. The equations of coupled motion of tall buildings are expressed first and the decoupling procedure using the modal superposition method is given. Then the uncoupled SDF equation in a generalized coordinate system is solved in the frequency domain based on random vibration theory. A more accurate method for quantifying intermodal cross correlation is analytically developed and a closed-form expression for the intermodal correlation coefficient considering the imaginary part of cross modal force spectra is given. By introducing spectral moments of the power spectral density (PSD) function in the frequency domain, an exact method for evaluating the intermodal correlation is also developed. In this exact method, the intermodal correlation coefficients can be given using a linear combination of modal response spectral moments related to any two mode-generalized single-degree-of-freedom (SDF) oscillators. Finally, using the aerodynamic wind loads derived from wind-tunnel based SMPSS test measurements, a full-scale 60-story building with 3D mode shapes and closely spaced frequencies is used to demonstrate the effectiveness and practicality of the proposed dynamic analysis framework and to highlight the effects of the cross correlations of modal responses on predicting the wind-induced acceleration of the building

## 3.2 Analysis of wind-induced coupled response

### 3.2.1 Equations of motion

Multi-story buildings with rigid floor diaphragms can be modeled by a lumped mass system having three degrees of freedom at each floor level (i.e., the $x$- and $y$-translations of the reference center and the $\theta$-rotation about a vertical axis through the reference center). Consider a building having $N$ stories as a $3N$-degree-of-freedom lumped mass system. The dynamic equilibrium of the building can be written as

$$\begin{bmatrix} \mathbf{M} & \mathbf{0} & \mathbf{0} \\ \mathbf{0} & \mathbf{M} & \mathbf{0} \\ \mathbf{0} & \mathbf{0} & \mathbf{I} \end{bmatrix}\begin{Bmatrix} \ddot{\mathbf{X}} \\ \ddot{\mathbf{Y}} \\ \ddot{\mathbf{\Theta}} \end{Bmatrix} + \begin{bmatrix} \mathbf{C}_{XX} & \mathbf{0} & \mathbf{C}_{X\Theta} \\ \mathbf{0} & \mathbf{C}_{YY} & \mathbf{C}_{Y\Theta} \\ \mathbf{C}_{X\Theta}^T & \mathbf{C}_{Y\Theta}^T & \mathbf{C}_{\Theta\Theta} \end{bmatrix}\begin{Bmatrix} \dot{\mathbf{X}} \\ \dot{\mathbf{Y}} \\ \dot{\mathbf{\Theta}} \end{Bmatrix} + \begin{bmatrix} \mathbf{K}_{XX} & \mathbf{0} & \mathbf{K}_{X\Theta} \\ \mathbf{0} & \mathbf{K}_{YY} & \mathbf{K}_{Y\Theta} \\ \mathbf{K}_{X\Theta}^T & \mathbf{K}_{Y\Theta}^T & \mathbf{K}_{\Theta\Theta} \end{bmatrix}\begin{Bmatrix} \mathbf{X} \\ \mathbf{Y} \\ \mathbf{\Theta} \end{Bmatrix} = \begin{Bmatrix} \mathbf{F}_X \\ \mathbf{F}_Y \\ \mathbf{F}_\Theta \end{Bmatrix} \quad (3.1)$$

where $\mathbf{X}=(x_1,x_2,...,x_N)^T$, $\mathbf{Y}=(y_1,y_2,...,y_N)^T$, $\mathbf{\Theta}=(\theta_1,\theta_2,...,\theta_N)^T$ are the displacement response sub-vectors; $\mathbf{M}=\text{diag}[m_i]$ represents the mass sub-matrix in which $m_i$ = the lumped mass at floor $i$; $\mathbf{I}=\text{diag}[I_i]$ denotes the mass moment of inertia matrix of the floor diaphragm about the vertical axis through the reference centre in which $I_i = m_i r_i^2$ and $r_i$ = the radius of gyration of each floor; $\mathbf{F}_X=(F_{x1},F_{x2},...,F_{xN})^T$, $\mathbf{F}_Y=(F_{y1},F_{y2},...,F_{yN})^T$, $\mathbf{F}_\Theta=(T_{\theta1},T_{\theta2},...,T_{\theta N})^T$ are the translational wind load sub-vectors in the $x$, $y$ directions, and the torsional wind moment about the vertical axis, respectively; $\mathbf{K}_{XX}$, $\mathbf{K}_{X\Theta}$, $\mathbf{K}_{YY}$, $\mathbf{K}_{Y\Theta}$, $\mathbf{K}_{\Theta\Theta}$ = the $N \times N$ stiffness sub-matrices of the building; and $\mathbf{C}_{XX}$, $\mathbf{C}_{X\Theta}$, $\mathbf{C}_{YY}$, $\mathbf{C}_{Y\Theta}$, $\mathbf{C}_{\Theta\Theta}$ = the $N \times N$ damping sub-matrices of the building.

Using the modal superposition method, the physical displacement response of a $N$-story building can be expressed in terms of the modal responses of $j=1, 2,..., n$ modes as

$$\begin{aligned} \mathbf{X} &= [\Phi_{jx}]\{q_j\} \\ \mathbf{Y} &= [\Phi_{jy}]\{q_j\} \\ \mathbf{\Theta} &= [\Phi_{j\theta}]\{q_j\} \end{aligned} \qquad (3.2)$$

where $q_j$=the $j$-th modal displacement; $[\Phi_{js}]$ denotes the $j$-th mode shape sub-matrix for the $s=x$, $y$, $\theta$ component. The mode shape matrix $\Phi$ can be written as

$$\Phi = \begin{bmatrix} \Phi_{1x} & \Phi_{2x} \cdots \Phi_{jx} \cdots & \Phi_{nx} \\ \Phi_{1y} & \Phi_{2y} \cdots \Phi_{jy} \cdots & \Phi_{ny} \\ \Phi_{1\theta} & \Phi_{2\theta} \cdots \Phi_{j\theta} \cdots & \Phi_{n\theta} \end{bmatrix} \qquad (3.3)$$

which can be determined from an eigenvalue analysis of the undamped free-vibration of the

49

structural system expressed in the form of

$$\begin{bmatrix} \mathbf{K}_{XX} & \mathbf{0} & \mathbf{K}_{X\Theta} \\ \mathbf{0} & \mathbf{K}_{YY} & \mathbf{K}_{Y\Theta} \\ \mathbf{K}_{X\Theta}^T & \mathbf{K}_{Y\Theta}^T & \mathbf{K}_{\Theta\Theta} \end{bmatrix} \begin{bmatrix} \Phi_x \\ \Phi_y \\ \Phi_\theta \end{bmatrix} = \begin{bmatrix} \mathbf{M} & \mathbf{0} & \mathbf{0} \\ \mathbf{0} & \mathbf{M} & \mathbf{0} \\ \mathbf{0} & \mathbf{0} & \mathbf{I} \end{bmatrix} \begin{bmatrix} \Phi_x \\ \Phi_y \\ \Phi_\theta \end{bmatrix} \begin{bmatrix} \omega_1^2 & 0 & 0 \\ 0 & \ddots & 0 \\ 0 & 0 & \omega_n^2 \end{bmatrix} \tag{3.4}$$

where $\omega_j$ =the $j$-th modal circular frequency.


### 3.2.2 Vibration analysis in frequency domain

Considering the orthogonality conditions of mode shapes and assuming classical damping, the dynamic equilibrium equation of motion from Eq. (3.1) can be transformed into a system of $j=1$, 2,..., $n$ uncoupled equations as

$$\begin{bmatrix} m_1 & 0 & 0 \\ 0 & \ddots & 0 \\ 0 & 0 & m_n \end{bmatrix} \begin{Bmatrix} \ddot{q}_1(t) \\ \vdots \\ \ddot{q}_n(t) \end{Bmatrix} + \begin{bmatrix} c_1 & 0 & 0 \\ 0 & \ddots & 0 \\ 0 & 0 & c_n \end{bmatrix} \begin{Bmatrix} \dot{q}_1(t) \\ \vdots \\ \dot{q}_n(t) \end{Bmatrix} + \begin{bmatrix} k_1 & 0 & 0 \\ 0 & \ddots & 0 \\ 0 & 0 & k_n \end{bmatrix} \begin{Bmatrix} q_1(t) \\ \vdots \\ q_n(t) \end{Bmatrix} = \begin{Bmatrix} Q_1(t) \\ \vdots \\ Q_1(t) \end{Bmatrix} \tag{3.5}$$

where the $j$-th mode generalized mass ($m_j$), damping ($c_j$), stiffness ($k_j$) and forces ($Q_j$) of the system can be given respectively as follows

$$m_j = \Phi_{jx}^T \mathbf{M} \Phi_{jx} + \Phi_{jy}^T \mathbf{M} \Phi_{jy} + \Phi_{j\theta}^T \mathbf{I} \Phi_{j\theta} \tag{3.6}$$

$$c_j = 2\xi_j m_j \omega_j \tag{3.7}$$

$$k_j = m_j \omega_j^2 \tag{3.8}$$

$$Q_j = \Phi_{jx}^T \mathbf{F_X} + \Phi_{jy}^T \mathbf{F_Y} + \Phi_{j\theta}^T \mathbf{F_\Theta} \tag{3.9}$$

Note that $\xi_j$ denotes the $j$-th modal damping ratio of the building.

Using the HFFB or SMPSS technique, aerodynamic wind loads on a building can be measured in a boundary layer wind tunnel, after which the dynamic response of the building can be calculated for any combination of the building's mass, stiffness and damping ratio. Based on random vibration theory, the spectral density matrix of the modal displacement response vector $(q_1, q_2, ..., q_n)$ can be calculated in the frequency domain as

$$[S_{q_j}(\omega)] = \mathbf{H}(\omega) \mathbf{S}_Q(\omega) \mathbf{H}^*(\omega), \qquad j = 1, 2, ..., n \tag{3.10}$$

where $\mathbf{S}_Q$=the matrix of the input modal wind load spectra which can be derived from the measured wind tunnel loading data time history; $\mathbf{H}(\omega) = diag[\dfrac{1}{m_j \omega_j^2} H_j(\omega)]$ =the matrix of the system

frequency response functions; and $\mathbf{H}^* =$ the complex conjugate of $\mathbf{H}$. The diagonal elements of the modal frequency response function in the matrix $\mathbf{H}$ can be expressed in a non-dimensional manner as a complex function as

$$H_j(\omega) = \frac{1}{1-(\omega/\omega_j)^2 + 2i\xi_j \omega/\omega_j}, \qquad j=1,2,...,n \qquad (3.11)$$

Generalized by mode shapes, the PSD of the modal wind forces can be related to the PSD functions of the measured aerodynamic wind forces as

$$\mathbf{S}_Q(\omega) = \mathbf{\Phi}^T [S_{F_{sl}}(\omega)] \mathbf{\Phi} \qquad (3.12)$$

where the sub-matrix $S_{F_{sl}}(\omega)$ is the PSD function of the measured aerodynamic force vector $\{F_X,$ $F_Y, F_\theta\}^T$ given in Eq. (3.1) and $s$, $l=x$, $y$, $\theta$ represent the three directional components of the aerodynamic wind forces.

Considering only the first three fundamental modes, Eq. (3.12) can be expanded into a long form as

$$\begin{bmatrix} S_{Q_{11}} & S_{Q_{12}} & S_{Q_{13}} \\ S_{Q_{21}} & S_{Q_{22}} & S_{Q_{23}} \\ S_{Q_{31}} & S_{Q_{32}} & S_{Q_{33}} \end{bmatrix} = \begin{bmatrix} \Phi_{1x} & \Phi_{1y} & \Phi_{1\theta} \\ \Phi_{2x} & \Phi_{2y} & \Phi_{2\theta} \\ \Phi_{3x} & \Phi_{3y} & \Phi_{3\theta} \end{bmatrix} \begin{bmatrix} S_{F_{xx}} & S_{F_{xy}} & S_{F_{x\theta}} \\ S_{F_{yx}} & S_{F_{yy}} & S_{F_{y\theta}} \\ S_{F_{\theta x}} & S_{F_{\theta y}} & S_{F_{\theta\theta}} \end{bmatrix} \begin{bmatrix} \Phi_{1x} & \Phi_{2x} & \Phi_{3x} \\ \Phi_{1y} & \Phi_{2y} & \Phi_{3y} \\ \Phi_{1\theta} & \Phi_{2\theta} & \Phi_{3\theta} \end{bmatrix} \qquad (3.13)$$

It is worth noting the following two special cases. If the mode shape matrix $\mathbf{\Phi}$ is diagonal, the system will become mechanically uncoupled. If the wind force spectral matrix $[S_{F_{sl}}]$ is strongly diagonal, the wind force action in different component directions will be deemed statistically uncoupled such that the cross correlations of wind loads acting in different component directions will then be negligible.

From the modal displacement response spectra given in Eq. (3.10), the PSD matrix of the physical displacement can be written as

$$[S_s(\omega)] = \mathbf{\Phi}[S_{q_j}(\omega)]\mathbf{\Phi}^T \qquad (3.14)$$

Substituting Eqs. (3.10) and (3.12) into Eq. (3.14), one gets the spectral density matrix of the respective displacement response vectors $\mathbf{X}(t)$, $\mathbf{Y}(t)$ and $\mathbf{\Theta}(t)$ as follows

$$[S_s(\omega)] = \mathbf{\Phi H}(\omega)\mathbf{\Phi}^T[S_{F_{sl}}(\omega)]\mathbf{\Phi H}^*(\omega)\mathbf{\Phi}^T, \quad (s,l=x,y,\theta) \qquad (3.15)$$

Using the, so-called, full spectral approach given in Eq. (3.15), the PSDs of the physical displacements considering all auto and cross modal correlations can be accurately determined. Considering that the PSDs of wind forces are one-sided spectra wherein $0 \le \omega \le \infty$, the mean square value of the system responses, in terms of displacements, velocities or accelerations can be calculated by integration in the frequency domain respectively as

$$[\sigma_s^2] = \int_0^\infty [S_s(\omega)]d\omega \qquad (3.16)$$

$$[\sigma_{\dot{s}}^2] = \int_0^\infty \omega^2 [S_s(\omega)]d\omega \qquad (3.17)$$

$$[\sigma_{\ddot{s}}^2] = \int_0^\infty \omega^4 [S_s(\omega)]d\omega \qquad (3.18)$$

The above sets of equations can be viewed as spectral moments associated with the PSD function $S_s(\omega)$.

Given with the $s$ component of $n$ number of 3D mode shapes of the building system at elevation $z$, i.e., $\{\phi_s(z)\}^T = \{\phi_{1s}(z), \phi_{2s}(z), ..., \phi_{ns}(z)\}$, the mean square value of the $s$-th component displacement response of the building system at elevation $z$ can be given as

$$\sigma_s^2 = \{\phi_s(z)\}^T \left[ \int_0^\infty \mathbf{H}(\omega)\mathbf{S}_Q(\omega)\mathbf{H}^*(\omega)d\omega \right] \{\phi_s(z)\}, \quad (s=x,y,\theta) \qquad (3.19)$$

In the form of modal combination after matrix manipulation, the $s$ component displacement response given in Eq. (3.19) can be written as

$$\sigma_s^2 = \sum_{j=1}^n \phi_{js}^2(z)\sigma_{q_{jj}}^2 + \sum_{j=1}^n \sum_{\substack{k=1 \\ j \ne k}}^n \phi_{js}(z)\phi_{ks}(z)C_{q_{jk}}, \quad (s=x,y,\theta) \qquad (3.20)$$

where $\sigma_{q_{jj}}^2$ represents the variance or mean square of the $j$-th modal displacement response and $C_{q_{jk}}$ denoting the covariance between the $j$-th and $k$-th modal displacement responses can be given as

52

$$C_{q_{jk}} = \frac{1}{m_j \omega_j^2 m_k \omega_k^2} \text{Re}\left[\int_0^\infty H_j(\omega) S_{Q_{jk}}(\omega) H_k^*(\omega) d\omega\right] \qquad (3.21)$$

in which Re represents the real operator. In the case of $j = k$, the covariance is reduced into the variance $\sigma_{q_{jj}}^2$, which can be written as

$$\sigma_{q_{jj}}^2 = \frac{\int_0^\infty |H_j(\omega)|^2 S_{Q_{jj}}(\omega) d\omega}{(m_j \omega_j^2)^2} \qquad (3.22)$$

Note that the first part of the right-hand-side of Eq. (3.20) indicates the contribution of all the modal auto-correlations and the second part represents the contribution of the intermodal cross-correlation taking in account both the mechanical and statistical coupling effects between any two different modes. While the mechanical coupling is reflected by 3D mode shapes, the statistical coupling can be measured by the, so called, intermodal correlation coefficient. Theoretically, by normalizing the covariance between two modes given in Eq. (3.21) by their associated root-mean-square modal responses $\sigma_{q_{jj}}$ and $\sigma_{q_{kk}}$ given in Eq. (3.22), the intermodal correlation coefficient $r_{jk}$ can be defined as

$$r_{jk} = \frac{C_{q_{jk}}}{\sigma_{q_{jj}} \sigma_{q_{kk}}} = \frac{\text{Re}[\int_0^\infty H_j(\omega) H_k^*(\omega) S_{Q_{jk}}(\omega) d\omega]}{\sqrt{\int_0^\infty H_j^2(\omega) S_{Q_{jj}}(\omega) d\omega}\sqrt{\int_0^\infty H_k^2(\omega) S_{Q_{kk}}(\omega) d\omega}} \qquad (3.23)$$

Using the intermodal correlation coefficient $r_{jk}$ given in Eq. (3.23), the mean square $s$ component of the displacement response of the building system can be rewritten as

$$\sigma_s^2 = \sum_{j=1}^n \phi_{js}^2(z)\sigma_{q_{jj}}^2 + \sum_{\substack{j=1 \\ j \neq k}}^n \sum_{k=1}^n \phi_{js}(z)\phi_{ks}(z) r_{jk}\sigma_{q_{jj}}\sigma_{q_{kk}}, \qquad (s = x, y, \theta) \quad (3.24)$$

Similarly, the $s$ component acceleration response can also be written by combining the modal acceleration response as

$$\sigma_{\ddot{s}}^2 = \sum_{j=1}^n \phi_{js}^2(z)\sigma_{\ddot{q}_{jj}}^2 + \sum_{\substack{j=1 \\ j \neq k}}^n \sum_{k=1}^n \phi_{js}(z)\phi_{ks}(z) r_{jk}\sigma_{\ddot{q}_{jj}}\sigma_{\ddot{q}_{kk}}, \qquad (s = x, y, \theta) \quad (3.25)$$

where the variance or mean square of the $j$-th modal acceleration response, $\sigma^2_{\ddot{q}_{jj}}$, can be given as

$$\sigma^2_{\ddot{q}_{jj}} = \frac{\int_0^\infty \omega^4 \left|H_j(\omega)\right|^2 S_{Q_{jj}}(\omega)d\omega}{(m_j\omega_j^2)^2} \qquad (3.26)$$

Assuming the generalized wind force spectrum as white noise, the mean square of the modal acceleration response can be approximated as

$$\sigma^2_{\ddot{q}_{jj}} \approx \frac{S_{Q_{jj}}(\omega_j)\int_0^\infty \omega^4 \left|H_j(\omega)\right|^2 d\omega}{(m_j\omega_j^2)^2} \qquad (3.27)$$

Since the squared modulus of the complex frequency response function $H_j(\omega)$ is a narrow band function that peaks at $\omega_j$, the 4-th order spectral moments of the squared modulus can then be simplified as

$$\int_0^\infty \omega^4 \left|H_j(\omega)\right|^2 d\omega \approx \omega_j^4 \int_0^\infty \left|H_j(\omega)\right|^2 d\omega \qquad (3.28)$$

Using the residual theorem, the integral of $\int_0^\infty \left|H_j(\omega)\right|^2 d\omega$ can be analytically expressed in terms of modal frequencies $\omega_j$ and the damping ratios $\xi_j$ as

$$\int_0^\infty \left|H_j(\omega)\right|^2 d\omega = \frac{\pi\omega_j}{4\xi_j} \qquad (3.29)$$

As a result of Eqs. (3.27), (3.28) and (3.29), the variance or mean square value of the $j$-th modal acceleration response can be expressed approximately in the form of an algebraic function as (Islam et al. 1992)

$$\sigma^2_{\ddot{q}_{jj}} \approx \frac{\pi\omega_j}{4\xi_j m_j^2} S_{Q_{jj}}(\omega_j) \qquad (3.30)$$

## 3.3   Estimation of intermodal correlation coefficient

In general, the covariance between the $j$-th and $k$-th modal displacement responses $C_{q_{jk}}$ involves the integral of two complex functions, i.e., $H_j$ being the complex frequency response function and $S_{Q_{jk}}$ the XPSD of generalized forces. By separating the real and imaginary parts of the two complex functions, the covariance given in Eq. (3.21), can be written in the following form

$$C_{q_{jk}} = \frac{\int_0^\infty \left\{ \text{Re}\left[ S_{Q_{jk}}(\omega) \right] \text{Re}\left[ H_j(\omega) H_k^*(\omega) \right] - \text{Im}\left[ S_{Q_{jk}}(\omega) \right] \text{Im}\left[ H_j(\omega) H_k^*(\omega) \right] \right\} d\omega}{m_j \omega_j^2 m_k \omega_k^2} \quad (3.31)$$

where Re denotes the real operator and Im indicates the imaginary operator. The real and imaginary parts of the product of modal frequency response functions can be given analytically as

$$\text{Re}[H_j(\omega) H_k^*(\omega)] = \frac{\left( \omega_j^2 - \omega^2 \right)\left( \omega_k^2 - \omega^2 \right) + 4\xi_j \xi_k \omega_j \omega_k \omega^2}{\left[ \left( \omega_j^2 - \omega^2 \right)^2 + 4\xi_j^2 \omega_j^2 \omega^2 \right]\left[ \left( \omega_k^2 - \omega^2 \right)^2 + 4\xi_k^2 \omega_k^2 \omega^2 \right]} \omega_j^2 \omega_k^2 \quad (3.32)$$

$$\text{Im}[H_j(\omega) H_k^*(\omega)] = \frac{2\omega_k \xi_k \left( \omega_j^2 - \omega^2 \right) - 2\omega_j \xi_j \left( \omega_k^2 - \omega^2 \right)}{\left[ \left( \omega_j^2 - \omega^2 \right)^2 + 4\xi_j^2 \omega_j^2 \omega^2 \right]\left[ \left( \omega_k^2 - \omega^2 \right)^2 + 4\xi_k^2 \omega_k^2 \omega^2 \right]} \omega \omega_j^2 \omega_k^2 \quad (3.33)$$

By neglecting the imaginary part of Eq. (3.31), the covariance of $j$-th and $k$-th modal responses in Eq. (3.31) can be simplified as

$$C_{q_{jk}} \approx \frac{\int_0^\infty \left\{ \text{Re}\left[ S_{Q_{jk}}(\omega) \right] \text{Re}\left[ H_j(\omega) H_k^*(\omega) \right] \right\} d\omega}{m_j \omega_j^2 m_k \omega_k^2} \quad (3.34)$$

Based on the simplified covariance response given in the above equation, the intermodal correlation coefficient $r_{jk}$ defined in Eq. (3.23) can be approximated as

$$r_{jk} \approx \frac{\int_0^\infty \left\{ \text{Re}\left[ S_{Q_{jk}}(\omega) \right] \text{Re}\left[ H_j(\omega) H_k^*(\omega) \right] \right\} d\omega}{\sqrt{\int_0^\infty H_j^2(\omega) S_{Q_{jj}}(\omega) d\omega} \sqrt{\int_0^\infty H_k^2(\omega) S_{Q_{kk}}(\omega) d\omega}} \quad (3.35)$$

Assuming that the wind-induced modal force spectra (i.e., $S_{Q_{jk}}, S_{Q_{jj}}$ and $S_{Q_{kk}}$) in Eq. (3.35) are white-noise spectra and considering the mean of the two modal frequencies, $\omega_j$ and $\omega_k$ as $\omega_{jk} = \dfrac{\omega_j + \omega_k}{2}$, the approximate intermodal correlation coefficient given in Eq. (3.35) can be further written as

$$r_{jk} \approx \frac{\text{Re}[S_{Q_{jk}}(\omega_{jk})]}{\sqrt{S_{Q_{jj}}(\omega_j) S_{Q_{kk}}(\omega_k)}} \times \frac{\int_0^\infty \text{Re}\left[ H_j(\omega) H_k^*(\omega) \right] d\omega}{\sqrt{\int_0^\infty H_j(\omega) H_j^*(\omega) d\omega \int_0^\infty H_k(\omega) H_k^*(\omega) d\omega}} \quad (3.36)$$

55

Using the residual theorem, the following integrals can be analytically expressed in terms of modal frequencies $\omega_j$, $\omega_k$ with corresponding respective damping ratios $\xi_j$, $\xi_k$ as

$$\int_0^\infty H_j(\omega)H_j^*(\omega)d\omega = \int_0^\infty \left|H_j(\omega)\right|^2 d\omega = \frac{\pi\omega_j}{4\xi_j} \qquad (3.37)$$

$$\int_0^\infty \text{Re}\left[H_j(\omega)H_k^*(\omega)\right]d\omega = \frac{2\pi(\xi_j\omega_j + \xi_k\omega_k)\omega_j^2\omega_k^2}{(\omega_j^2 - \omega_k^2)^2 + 4(\xi_j\omega_j + \xi_k\omega_k)(\xi_k\omega_k\omega_j^2 + \xi_j\omega_j\omega_k^2)} \qquad (3.38)$$

Substituting Eqs. (3.37) and (3.38) into (3.36), the intermodal correlation coefficient $r_{jk}$ can then be further simplified into an algebraic form as (Chen and Kareem, 2005a)

$$r_{jk} \approx \frac{\text{Re}[S_{Q_{jk}}(\omega_{jk})]}{\sqrt{S_{Q_{jj}}(\omega_j)S_{Q_{kk}}(\omega_k)}} \times \rho_{jk} \qquad (3.39)$$

where the coefficient $\rho_{jk}$ is the well known complete quadratic combination (CQC) factor for modal response superposition first given by Der Kiureghian (1981) for earthquake engineering applications as

$$\rho_{jk} = \frac{8\sqrt{\xi_j\xi_k\omega_j\omega_k}(\xi_j\omega_j + \xi_k\omega_k)\omega_j\omega_k}{(\omega_j^2 - \omega_k^2)^2 + 4\xi_j\xi_k\omega_j\omega_k(\omega_j^2 + \omega_k^2) + 4(\xi_j^2 + \xi_k^2)\omega_j^2\omega_k^2} \qquad (3.40)$$

In earthquake engineering, a seismic action on a building comes generally from the base level of the building, the modal earthquake force spectra for various modes of vibration of the building must, therefore, be fully coherent such that $S_{Q_{jk}} = \sqrt{S_{Q_{jj}}S_{Q_{kk}}}$. As a result, the earthquake force spectra ratio, $\dfrac{\text{Re}[S_{Q_{jk}}(\omega_{jk})]}{\sqrt{S_{Q_{jj}}(\omega_j)S_{Q_{kk}}(\omega_k)}}$, given in Eq. (3.39), can be reduced to 1 and the corresponding intermodal correlation coefficient for a single-source seismic action on a building can also be reduced to the traditional CQC factor as follows

$$r_{jk} \approx \rho_{jk} \qquad (3.41)$$

However in the situation of wind loading actions which can be regarded as multi-point loading excitations, the modal wind force spectra ratio given in Eq. (3.39) may not generally be equal to 1 due to the spatial and temporal nature of wind. When considering the partial correlation effects of the spatial distribution of stochastic wind forces, the wind force spectra ratio may vary within the range as $-1 \leq \dfrac{\text{Re}[S_{Q_{jk}}(\omega_{jk})]}{\sqrt{S_{Q_{jj}}(\omega_j)S_{Q_{kk}}(\omega_k)}} \leq 1$. For ease of reference, the CQC method using the approximate

intermodal correlation coefficient of Eq. (3.40), in which the generalized forces are assumed to be fully correlated as in the case of a single-source earthquake excitation, is regarded as the traditional CQC (TCQC) rule. Similar to the situation of a tall building under multi-point stationary random wind excitations, the incoherency of the spatially varying multiple earthquake excitations acting on a multisupport-structural system (e.g., long span bridge) is found to have a significant influence on the dynamic response (Heredia-Zavoni and Vanmarcke 1994; Allam and Datta 1999, 2004).

To further improve the accuracy of the intermodal correlation coefficient $r_{jk}$, the imaginary parts of both the complex frequency functions and the XPSD function between two modal forces $Q_j(t)$ and $Q_k(t)$ can be included in the calculation of the covariance as given in Eq. (3.31). By normalizing the covariance response given in Eq. (3.31) by the two associated root mean square (RMS) modal responses, the intermodal correlation coefficient $r_{jk}$ can be accurately expressed as

$$r_{jk} = \frac{\int_0^\infty \left\{ \text{Re}\left[ S_{Q_{jk}}(\omega) \right] \text{Re}\left[ H_j(\omega)H_k^*(\omega) \right] - \text{Im}\left[ S_{Q_{jk}}(\omega) \right] \text{Im}\left[ H_j(\omega)H_k^*(\omega) \right] \right\} d\omega}{\sqrt{\int\limits_0^\infty H_j^2(\omega)S_{Q_{jj}}(\omega)d\omega}\sqrt{\int\limits_0^\infty H_k^2(\omega)S_{Q_{kk}}(\omega)d\omega}} \quad (3.42)$$

Again, using the residual theorem, the integral $\int_0^\infty \text{Im}\left[ H_j(\omega)H_k^*(\omega) \right]\dfrac{d\omega}{\omega}$ with the poles of the

integrand at $\pm\omega_j\sqrt{(1-\xi_j^2)}+i\xi_j\omega_j$ and $\pm\omega_k\sqrt{(1-\xi_k^2)}+i\xi_k\omega_k$ can be solved analytically in a closed-form expression as

$$\int_0^\infty \text{Im}\left[ H_j(\omega)H_k^*(\omega) \right]\frac{d\omega}{\omega} = \frac{\pi}{2}\frac{\omega_j^4 - \omega_k^4 + 4(\xi_j\omega_j + \xi_k\omega_k)(\xi_k\omega_k\omega_j^2 - \xi_j\omega_j\omega_k^2)}{(\omega_j^2 - \omega_k^2)^2 + 4(\xi_j\omega_j + \xi_k\omega_k)(\xi_k\omega_k\omega_j^2 + \xi_j\omega_j\omega_k^2)} \quad (3.43)$$

Assuming the cross spectral density function $S_{Q_{jk}}$ as white noise and separating the real part and imaginary part of the complex integrals, the intermodal correlation coefficient $r_{jk}$ can be rewritten analytically as

$$r_{jk} \approx \frac{\text{Re}[S_{Q_{jk}}(\omega_{jk})]}{\sqrt{S_{Q_{jj}}(\omega_j)S_{Q_{kk}}(\omega_k)}} \times \rho_{jk} - \frac{\text{Im}[S_{Q_{jk}}(\omega_{jk})]}{\sqrt{S_{Q_{jj}}(\omega_j)S_{Q_{kk}}(\omega_k)}} \times \rho_{jk}^{(I)} \quad (3.44)$$

where the traditional CQC factor $\rho_{jk}$ is given in Eq. (3.40) and the new imaginary CQC factor

$\rho_{jk}^{(I)}$ can be approximated into a form of algebraic expression as:

$$\rho_{jk}^{(I)} = \frac{\int_0^\infty \mathrm{Im}\left[H_j(\omega)H_k^*(\omega)\right]d\omega}{\sqrt{\int_0^\infty H_j^2(\omega)d\omega}\sqrt{\int_0^\infty H_k^2(\omega)d\omega}} \approx \frac{\omega_{jk}\int_0^\infty \mathrm{Im}\left[H_j(\omega)H_k^*(\omega)\right]\dfrac{d\omega}{\omega}}{\sqrt{\int_0^\infty H_j^2(\omega)d\omega}\sqrt{\int_0^\infty H_k^2(\omega)d\omega}}$$

$$\approx 2\omega_{jk}\sqrt{\frac{\xi_j\xi_k}{\omega_j\omega_k}}\frac{\omega_j^4 - \omega_k^4 + 4(\xi_j\omega_j + \xi_k\omega_k)(\xi_k\omega_k\omega_j^2 - \xi_j\omega_j\omega_k^2)}{(\omega_j^2 - \omega_k^2)^2 + 4(\xi_j\omega_j + \xi_k\omega_k)(\xi_k\omega_k\omega_j^2 + \xi_j\omega_j\omega_k^2)} \tag{3.45}$$

The imaginary CQC factor $\rho_{jk}^{(I)}$, representing the imaginary part of the intermodal correlation of two modal responses, can be plotted as a function of the ratio of $\omega_j / \omega_k$ for two given values of the corresponding damping ratios $\xi_j$ and $\xi_k$. As shown in Figure 3.1, the coefficient $\rho_{jk}^{(I)}$ from Eq. (3.45) and the traditional CQC factor from Eq. (3.40) are depicted for the cases of $\xi_j, \xi_k = 2\%$ and $\xi_j, \xi_k = 5\%$. It can be observed that the traditional CQC factor $\rho_{jk}$, which is a symmetric even function about the frequency ratio of 1, diminishes rapidly as the two frequencies depart from each other. However, the imaginary CQC factor $\rho_{jk}^{(I)}$, which is an asymmetric odd function about the frequency ratio of 1, does not diminish as quickly as the frequency ratio moves far away from the value of one.

Using the concept of spectral moments, it is also possible to evaluate numerically the intermodal correlation coefficient without any underlying assumptions and simplifications (i.e. without the use of assumptions about white noise generalized wind force spectra and narrow band frequency response functions). For this purpose, it is necessary to expand the real and imaginary parts of the product of modal frequency response functions in the form of a polynomial function of the circular frequency, respectively

$$\mathrm{Re}[H_j(\omega)H_k^*(\omega)] = \left|H_j(\omega)\right|^2 \left|H_k(\omega)\right|^2 \left[1 + \gamma_{jk}\omega^2 + \varepsilon_{jk}\omega^4\right] \tag{3.46}$$

$$\mathrm{Im}[H_j(\omega)H_k^*(\omega)] = \left|H_j(\omega)\right|^2 \left|H_k(\omega)\right|^2 \left[\alpha_{jk}\omega + \beta_{jk}\omega^3\right] \tag{3.47}$$

where

$$\gamma_{jk} = (4\xi_j\xi_k\omega_j\omega_k - \omega_j^2 - \omega_k^2)/(\omega_j^2\omega_j^2) \tag{3.48}$$

$$\varepsilon_{jk} = 1/(\omega_j^2\omega_k^2) \tag{3.49}$$

$$\alpha_{jk} = 2(\xi_k \omega_j - \xi_j \omega_k)/(\omega_j \omega_k) \tag{3.50}$$

$$\beta_{jk} = 2(\xi_j \omega_j - \xi_k \omega_k)/(\omega_j^2 \omega_k^2) \tag{3.51}$$

$$\left|H_j(\omega)\right|^2 = \frac{1}{\left[1-\left(\omega/\omega_j\right)^2\right]^2 + \left(2\xi_j \omega/\omega_j\right)^2} \tag{3.52}$$

Substituting Eq. (3.46) and (3.47) into Eq. (3.31) and then the resulting equation into Eq. (3.23), one obtains the exact intermodal correlation coefficient as follows

$$r_{jk} = \frac{1}{\sqrt{\lambda_{jj}\lambda_{kk}}}\left[\lambda_{0,jk} + \gamma_{jk}\lambda_{2,jk} + \varepsilon_{jk}\lambda_{4,jk} - \alpha_{jk}\lambda_{1,jk} - \beta_{jk}\lambda_{3,jk}\right] \tag{3.53}$$

where the spectral moments are given as

$$\lambda_{jj} = \int_0^\infty \left|H_j(\omega)\right|^2 S_{Q_{jj}}(\omega)d\omega \tag{3.54}$$

$$\lambda_{m,jk} = \begin{cases} \int_0^\infty \omega^m \left|H_j(\omega)\right|^2 \left|H_k(\omega)\right|^2 \mathrm{Re}\left[S_{Q_{jk}}(\omega)\right]d\omega, \ m=0,2,4 \\ \int_0^\infty \omega^m \left|H_j(\omega)\right|^2 \left|H_k(\omega)\right|^2 \mathrm{Im}\left[S_{Q_{jk}}(\omega)\right]d\omega, \ m=1,3 \end{cases} \tag{3.55}$$

in which $\lambda_{jj}$ indicates the spectral moments of the response of a single-degree oscillator having frequency $\omega_j$ and damping ratio $\xi_j$, while $\lambda_{m,jk}$ represents the cross-spectral moments associated with the cross modal spectrum $S_{Q_{jk}}$ of modes $j$ and $k$.

Without the assumption of white-noise spectra and the narrow band modal frequency response functions, the exact value of the intermodal correlation coefficient can be calculated from Eq. (3.53) by numerical integration of the spectral moments given in Eqs. (3.54) and (3.55). Once the exact value of the intermodal correlation efficient is obtained from Eq. (3.53), the corresponding modal combination using such intermodal correlation coefficient can be deemed as the exact complete quadratic combination (ECQC) method. For the sake of comparison, the complete quadratic combination method using the refined correlation coefficient formula of Eq. (3.44), which assumes white-noise wind spectra, but takes into account more accurately the partial correlation of wind loads and the quad-spectra (imaginary parts of XPSD) of generalized wind forces, can be called the accurate complete quadratic combination (ACQC) method. The CQC method using the simplified intermodal correlation coefficient of Eq. (3.40), which has been derived with the assumptions of the white-noise wind spectra and full coherence between any two modal force spectra, is regarded as

59

the traditional CQC (TCQC) method.

## 3.4   Illustrative example

A 60-story hybrid steel and concrete building as shown in Figure 3.2 was used to test the proposed framework for wind-induced dynamic response analysis. With a story height of 4 m, the building has an overall height of 240 m and a rectangular floor plan dimension of 24 m by 72 m. The building has an outrigger braced system, which consists of an eccentric concrete core connected to the exterior braced steel frame by two levels of steel outrigger trusses. Since the concrete core of the building is located eccentrically, the building is expected to exhibit significant coupled lateral-torsional effects under wind excitations.

A wind tunnel test was carried out at the CLP Power Wind/Wave Tunnel Facility (WWTF) of The Hong Kong University of Science and Technology (Tse et al., 2007). Aerodynamic wind forces acting on the building were measured by the SMPSS technique using a 1:400 scale rigid model. For this building example, two specific wind load cases were considered. One was the 0-degree wind perpendicular to the wide face acting in the short direction (i.e. along the Y-axis) of the building; another one was the 90-degree wind perpendicular to the narrow face acting in the long direction (i.e. along the X-axis). Ten-year return period mean wind speed of 35m/s at the top of the building was used to calculate wind-induced acceleration responses. The measured local pressure coefficients were converted into wind force by employing the following equation

$$F(t) = 0.5\rho\bar{U}^2 C_p(t)\Delta A \tag{3.56}$$

where $\rho$ is the air density; $\bar{U}$ is the reference mean wind speed at that local position; $C_p(t)$ is the pressure coefficient time history recorded from the wind tunnel test; $\Delta A$ is the tributary area of each pressure tap. Once the collective external wind forces of each floor level were calculated using Eq. (3.56) and combined with the mode shapes of the building, the modal wind forces were then determined using Eq. (3.9). The PSD functions of the modal forces were then computed through finite Fourier transformation and are shown in Figs. 3.3 and 3.4.

The initial element sizes of the structural system were established based on the code-specified strength design criteria for ensuring minimum safety requirements for the building. The initial dynamic properties of the building (i.e. the natural frequencies and mode shapes) were determined by eigenvalue analysis. The first three fundamental frequencies of the initial building system were 0.231 Hz (swaying primarily in the short direction of the building), 0.268 Hz (mainly torsional

60

vibration) and 0.273Hz (swaying primarily in the long direction). The first three fundamental mode shapes referenced about the geometric center of each floor along the height of the building are depicted in Figure 3.5. It should be noted that the torsional mode shapes were transformed into swaying translation by multiplying the angle of twist with the corresponding radius of gyration of each floor. Due to the asymmetric structural layout of the building, the first three modes resulted in 3D lateral-torsional shapes with one dominating component in each of the three modes. For the first three modes, a modal damping ratio of 2% was assumed.

Table 3.1 presents the RMS modal acceleration responses of the building computed using Eq. (3.30). Although the building has an elongated rectangular plan, the three modal accelerations due to the incident 90 degree wind are found to be larger than those due to the incident 0 degree wind. The larger modal accelerations associated with the incident 90 degree wind were mainly caused by the more significant vortex shedding effect found in the short direction of the building. For the incident wind angle of 90 degree with wind normal to the narrow face of the building, flow separations were first found at the leading corners of the windward short face of the building, then followed downstream by flow reattachments and finally, once again, by flow separations at the leeward edges of the building. The vortex shedding processes resulted in a higher level of wind-induced dynamic loads in the crosswind direction (short direction) of the building, leading to a more significant mode 1 acceleration response than that induced by the incident 0 degree wind. The mode 2 acceleration response due to the 90 degree wind could be attributed to the imbalanced flow fluctuations causing uneven pressure distribution and larger torsional loading effects on the building than that associated with the 0 degree wind. For mode 3, as a swaying mode primarily in the long direction but also accompanied with significant mechanical coupling components in the short direction, apparently the crosswind vibration along the short direction of the building due to the 90 degree wind significantly influenced the mode 3 acceleration, resulting in a larger modal acceleration due to the action of 90 degree wind.

The intermodal correlation coefficients as shown in Table 3.2 used for the three combination methods were calculated from Eq. (3.40) for TCQC, Eq. (3.44) for ACQC and Eq. (3.53) for ECQC, respectively. It is evident that the intermodal correlation coefficients used for TCQC were independent of the wind loading directions and were always positive. However, the intermodal correlation coefficients used for both the ACQC and ECQC methods were dependent on the incident wind directions and they could take on either positive or negative values. As shown in Table 3.2, there exist significant differences between the values of the intermodal correlation coefficients predicted by the TCQC and ECQC formulas. The differences were mainly due to the fact that the TCQC formula neglected the spectral characteristics of wind loads. For the two closely

61

spaced modes 2 and 3, the correlation coefficient $r_{23}$ was found to be equal to 0.8241 using TCQC, which was over-estimated by 28% and 137% compared with the exact values predicted by ECQC under the 90 degree wind and the 0 degree wind, respectively. When the mode 1 frequency was separated further from those of modes 2 and 3, negative values of $r_{12}$ and $r_{13}$ were found by both ACQC and ECQC. The negative correlations are attributed to the two associated vibration modes being out of phase. As shown in Figure 3.1, when the frequency ratio moves far away from unity, the imaginary CQC factor $\rho_{jk}^{(I)}$ becomes more dominant than the traditional CQC factor $\rho_{jk}$. This implies that the effects of the quad-spectra (Imaginary part) of XPSD of generalized wind forces become more significant than that of the co-spectra when considering two separated modes. Hence for $r_{12}$ and $r_{13}$, in terms of magnitude, TCQC under-estimated the intermodal correlation when compared with ECQC since significant effects of the quad-spectra of XPSD were neglected by TCQC.

Table 3.3 shows the results of the RMS acceleration response at the top level of the building. Based on the modal acceleration results, the physical acceleration responses were obtained by combining the modal acceleration results using the traditional CQC (TCQC), the accurate CQC (ACQC) and the exact CQC (ECQC) combination methods. It should be noted that the torsional accelerations given in Table 3.3 have been expressed as translational accelerations at the corner of the top level of the building. Since the building was weaker in its short direction (Y-axis), the RMS acceleration response along the Y-axis was always larger than that in the long direction (X-axis) even when the building was subjected to the 90 degree X-direction wind. Due to the asymmetric structural layout of the building, significant torsional accelerations were found such that the translational acceleration at the corner of the top level induced by the torsional effect was even larger than the Y-direction swaying acceleration at the center of the top level.

Some noticeable differences between the component acceleration results of the TCQC and ECQC methods were found in Table 3.3. When compared to the acceleration results of ECQC, the TCQC method resulted in large differences in terms of percentage errors ranging from -39.8% to 32.6%. The components of acceleration computed using the ACQC method were more comparable to that obtained by the ECQC method with a percentage difference ranging from -1.3% to 8.3%. Evidently, the ACQC method formulated here yields very encouraging results for the prediction of acceleration response, and agrees well with the acceleration values predicted by the ECQC method. These results clearly demonstrate the importance of considering the partially correlated characteristics of wind loads and the quad-spectra of XPSD of generalized wind forces for accurate predictions of wind-induced response.

The relationships between the intermodal correlation coefficient and the modal frequency ratio is depicted in Figs. 3.6 (a, b, c) for the 0 degree wind loading case and Figure 3.7 (a, b, c) for the 90 degree wind loading case, respectively. Assuming that the modal force spectra and frequency value of mode 2 is fixed, the intermodal correlation coefficient $r_{12}$ can be evaluated for varying values in the frequency of mode 1. Similarly, by fixing the frequency value of mode 3, the intermodal correlation coefficients $r_{13}$ and $r_{23}$ can be obtained by varying frequency values of modes 1 and 2, respectively. In Figs. 3.6 (a, b, c) and 3.7 (a, b, c), the intermodal correlation coefficients are presented as functions of the modal frequency ratio using the TCQC formula of Eq. (3.40), the ACQC formula of Eq. (3.44) and the ECQC formula of Eq. (3.53) for the three corresponding modal combination methods, respectively.

It can be observed From Figs. 3.6 and 3.7 that at the lower frequency ratio range (i.e. when two modal frequencies are well separated), say lower than a value of 0.85, the intermodal correlation coefficients calculated from the TCQC formula of Eq. (3.40)were always very small positive values (close to zero) and were independent of the incident wind directions. However, the exact intermodal correlation coefficients for $r_{12}$ and $r_{13}$ calculated by the ACQC formula of Eq. (3.44) or the ECQC formula of Eq. (3.53) were significant and had negative values over the whole lower frequency ratio range as shown in both Figs. 3.6(a, c) and 3.7(a, c). The negative correlation could be attributed to the out of phase phenomena between two separated vibration modes. Generally speaking, in the lower range of modal frequency ratio, the variation of correlation is quite steady as shown in Figs. 3.6 and 3.7.

When two modal frequencies are more closely spaced, say with a modal frequency ratio larger than 0.85, the values of the intermodal correlation coefficients change more rapidly. In general, the TCQC formula overestimates the intermodal correlation at the higher modal frequency range. The TCQC formula based on the assumption of fully coherent modal force spectra does not reflect the partially correlated characteristic of wind loads, thus leading to the overestimation of correlation. On the other hand, the correlation coefficients for any two modes predicted by the ACQC formula agree well with the exact value of the intermodal correlation calculated by the ECQC formula as shown in Figs. 3.6 and 3.7 for both the 0 degree and 90 degree wind cases. This close agreement validates the accuracy of the closed-form analytical ACQC formula of Eq. (3.44) for evaluating intermodal correlation coefficients.

For the correlation coefficients evaluated by the ACQC or ECQC formula, two patterns of variation at higher frequency ratio range can be identified. One is that the intermodal correlations vary monotonically with the change of modal frequency ratio. While the correlation coefficient $r_{23}$, as shown in Figure 3.7(b), increases with the modal frequency ratio, the correlation coefficients $r_{12}$ in Figure 3.6(a) and $r_{13}$ in Figure 3.6(c) decrease with the modal frequency ratio. Such monotonically increasing or decreasing variations are due to the fact that the contributions to the intermodal correlation are dominated by the co-spectra of XPSD. As shown in Eq. (3.44), when the co-spectra term is more dominant than the quad-spectra term, the intermodal correlation coefficient can be proportionally related to the traditional CQC factor. In fact, the traditional CQC factor curves shown in Figure 1 clearly reflect the monotonic variation trend of the intermodal correlation in the higher modal frequency ratio range.

The second variation pattern in the higher modal frequency ratio range can be observed in Figure 3.6(b) for $r_{23}$, Figure 3.7(a) for $r_{12}$ and Figure 3.7(c) for $r_{13}$, where the intermodal correlation coefficients firstly rise and then fall rapidly with a change of curvature. Such results indicate that these intermodal correlation coefficients are indeed more dominated by the quad-spectra of XPSD in the higher modal frequency ratio range. The change of curvatures in the intermodal correlation coefficient is mainly attributable to the characteristics of the imaginary CQC factor as shown in Figure 3.1.


## 3.5  Summary

This chapter presents a dynamic analysis framework for predicting the wind-induced response of tall buildings with 3D mode shapes. The cross-correlation of modal responses under spatiotemporally varying dynamic wind loads on buildings has been investigated in detail. Three different formulae for determining the intermodal correlation coefficient with various levels of accuracy have been established. Corresponding to these three different formulae for the intermodal correlation, three modal combination methods, namely the traditional CQC (TCQC), the accurate CQC (ACQC) and the exact CQC (ECQC) methods, have been developed for predicting wind-induced dynamic responses of tall buildings.

A prototype 60-story asymmetric building example has been utilized to illustrate the use of the three combination methods for assessing the cross-correlation of modal responses of the building to determine wind-induced lateral-torsional motions. Results have indicated that the application of the traditional CQC factor without using the information of wind load spectra may result in significant errors in the evaluation of intermodal correlation coefficients, particularly when two modal frequencies are closely spaced. The wind-induced acceleration response determined using the traditional CQC combination method may be significantly over-estimated or at times under-estimated, especially for buildings with closely spaced modal frequencies and 3D mode shapes.

The analytical closed form ACQC formula makes use of the information of both the real and imaginary parts of the cross power spectral density of generalized wind forces, providing accurate results of the intermodal correlation coefficients which are comparable to that of the ECQC formula. The acceleration results of the 60-story building example predicted by the ACQC method agree well with the results calculated by the ECQC method. Although the ACQC method is slightly less accurate than the ECQC method, which requires the use of more cumbersome numerical integration, the ACQC method in the analytical closed-form expression is more convenient for practical use and computationally more efficient.

Table 3.1    Modal RMS acceleration response of the 60-story hybrid building

| Wind cases | Modal acceleration (milli-g $\approx 10^{-2}$ m/s$^2$) | | |
| --- | --- | --- | --- |
| | Mode 1 | Mode 2 | Mode 3 |
| 0 degree wind | 10.71 | 7.44 | 5.72 |
| 90 degree wind | 12.85 | 12.38 | 9.53 |

Table 3.2    Intermodal correlation coefficients for the 60-story hybrid building

| Wind cases | Formula used | Intermodal correlation coefficients | | |
| --- | --- | --- | --- | --- |
| | | $r_{12}$ | $r_{23}$ | $r_{13}$ |
| 0 degree wind | TCQC | 0.0652 | 0.8241 | 0.0524 |
| 0 degree wind | ACQC | -0.3153 | 0.3615 | -0.1970 |
| 0 degree wind | ECQC | -0.3352 | 0.3473 | -0.1796 |
| 90 degree wind | TCQC | 0.0652 | 0.8241 | 0.0524 |
| 90 degree wind | ACQC | -0.0664 | 0.5645 | -0.0506 |
| 90 degree wind | ECQC | -0.0755 | 0.6428 | -0.0785 |

65

**Table 3.3   RMS acceleration response at the top of the 60-story hybrid building**

| Wind cases | Combination method | Acceleration component (milli-g $\approx 10^{-2}$ m/s$^2$) | | |
|---|---|---|---|---|
| | | X | Y | Torsion×37.95m |
| 0 degree wind | TCQC | 3.02 (-39.8%) | 11.43 (-11.7%) | 14.25 (32.6%) |
| 0 degree wind | ACQC | 4.97 (-0.9%) | 12.90 (-0.3%) | 10.85 (1.0%) |
| 0 degree wind | ECQC | 5.02 | 12.94 | 10.74 |
| 90 degree wind | TCQC | 5.04 (-22.7%) | 15.13 (-5.4%) | 18.08 (8.6%) |
| 90 degree wind | ACQC | 7.05 (8.3%) | 15.78 (-1.3%) | 16.44 (-1.2%) |
| 90 degree wind | ECQC | 6.51 | 15.99 | 16.65 |

Note: The numbers in brackets represent the percentage errors of acceleration component values calculated by TCQC or ACQC as compared with the exact value obtained by ECQC.

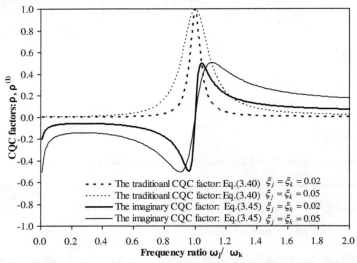

**Figure 3.1   The traditional CQC factor $\rho_{jk}$ and the imaginary CQC factor $\rho_{jk}^{(I)}$**

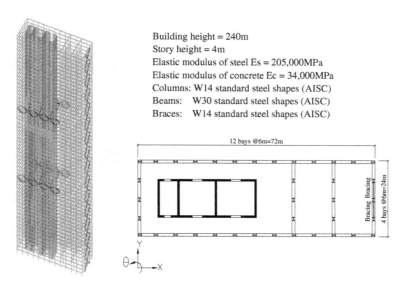

Building height = 240m
Story height = 4m
Elastic modulus of steel Es = 205,000MPa
Elastic modulus of concrete Ec = 34,000MPa
Columns: W14 standard steel shapes (AISC)
Beams:   W30 standard steel shapes (AISC)
Braces:  W14 standard steel shapes (AISC)

**Figure 3.2   A 60-story hybrid building with 2-story height outriggers**

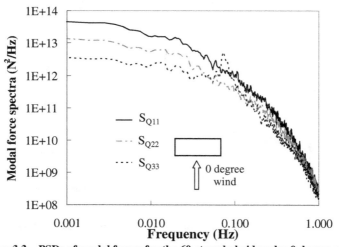

**Figure 3.3   PSDs of modal forces for the 60-story hybrid under 0 degree wind**

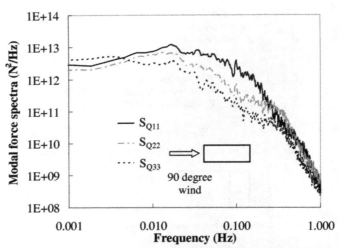

**Figure 3.4    PSDs of modal forces for the 60-story hybrid under 90 degree wind**

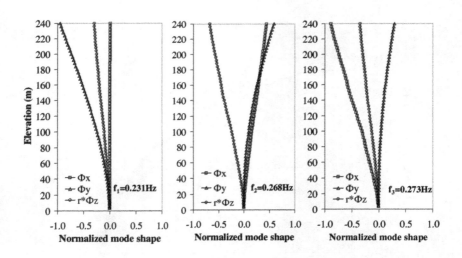

**Figure 3.5    3D mode shapes of the 60-story hybrid building**

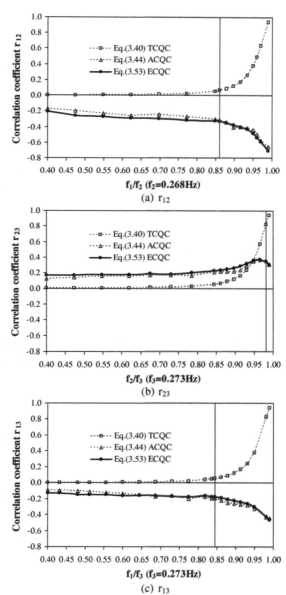

(a) $r_{12}$

(b) $r_{23}$

(c) $r_{13}$

Figure 3.6    Intermodal correlations due to the 0 degree wind

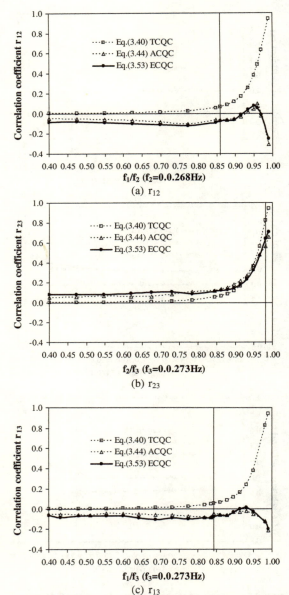

(a) $r_{12}$

(b) $r_{23}$

(c) $r_{13}$

Figure 3.7    Intermodal correlations due to the 90 degree wind

70

# CHAPTER 4    Wind Load Updating and Drift Design Optimization of Tall Buildings

## 4.1    Introduction

Modern wind-sensitive tall buildings are prone to dynamic serviceability problems. Although marked improvements have been made in many wind codes and standards for providing reasonable estimates of wind-induced structural loads on isolated buildings, current codes cannot account for effects of building shapes significantly different than simple rectangular geometries, nor for the interference effects caused by the surrounding structures in a complex terrain. Furthermore, most codes provide little guidance on accurately determining the crosswind forces induced by pressure fluctuations on the sides faces when wind flows around a building, nor can they address the wind-induced torsional effects resulting from an imbalance in the instantaneous pressure distributions on the building surfaces. Torsional twisting effects, resulting from an imbalance in the instantaneous pressure distribution on the building surface, are further amplified in tall asymmetric buildings by the presence of significant eccentricities between the center of stiffness of the structural system and the center of wind forces on the building.

In the past few decades, wind tunnels have emerged as an indispensable empirical tool for simulating natural winds and determining wind forces on structures using small-scale models in the simulated air flow (Cermak 2003). Boundary layer wind tunnel model testing has become the most reliable method for the determination of wind-induced structural loads and responses of tall buildings. With the development of advanced data acquisition and processing systems, accurate aerodynamic model testing has been made possible by the widely accepted high-frequency force balance (HFFB) techniques and more recently also by using the synchronous multi-pressure sensing system (SMPSS) technique. The SMPSS technique allows direct computation of mode generalized forces for any number of modes of vibration of the building with linear or nonlinear 3D modes. Both of these aerodynamic testing techniques have the advantages of utilizing simple lightweight but relatively rigid building models. Once the aerodynamic wind load spectra derived from wind tunnel testing are obtained, the equivalent static wind loads (ESWLs) can then be computed given the dynamic properties of the tested building. As long as the geometric configuration of the building does not change, the same set of aerodynamic load spectra can be used to estimate the wind-induced loads and responses without the need for further wind tunnel tests.

While wind tunnel techniques can provide accurate predictions of the structural loads for a building,

71

the current practice normally regards the wind tunnel derived ESWLs as similar to the code specified wind loads that are kept unchanged throughout the entire design synthesis process. It is not common to realize the fact that the wind-induced equivalent static loads on a building are indeed related to the dynamic properties (such as the effective stiffness, mass and damping) of a building. During the design process of a building, the structural system of the building is often modified, leading to a considerable difference in the structural dynamic properties of the building than the initial conditions of the building assumed at the time of the wind tunnel test. Depending on whether the structural system is weaken or stiffened, the wind-induced structural loads on a building can be increased or reduced as a function of the frequency of the building.

Figure 4.1 shows the spectra of the alongwind, crosswind and torsional base moments of typical square and rectangular tall buildings derived from a series of wind tunnel tests (Zhou, Kijewski and Kareem 2003). Such aerodynamic force spectra demonstrate peak values at a critical reduced frequency $f_C$ as shown in Figure 4.2 around or below 0.1. As the fundamental frequency of tall buildings is typically between 0.1 and 1 Hz, such structures are generally affected by the high reduced frequency range (i.e. $fB/U>0.1$, where $f$ = natural frequency, $B$ =building width normal to the oncoming wind, $U$ = wind speed at the top of the building), characterizing the descending part of the power spectra. As observed from Figure 4.1, the power spectral density of wind force reduces with increasing frequency, indicating that a reduction in the wind loads can be achieved by increasing the building's stiffness. Imagine that the computed reduced frequency of the initial trial design of a building at the time of a wind tunnel test is $f_o$ as shown in Figure 4.2. If the structural system of the building is stiffened during the design process, the frequency of the building will then be increased, causing also a subsequent reduction in the magnitude of the wind loads for the building.

Although automated optimization techniques are now made available for optimizing the lateral stiffness of tall building structures subject to static drift and dynamic frequency constraints (Chan 1997, 1998, 2001), the stiffness design optimization is normally carried out with a fixed set of wind loading conditions. Since the optimization techniques are generally applied without a proper update of the ESWLs, the stiffness enhancement achieved by such optimization techniques may likely result in an over-conservative design in which the structural load is probably overestimated. In order to capture accurately the wind induced structural loads on a building during the design optimization process and to achieve a cost efficient structural design solution, a computer based technique is developed with the aim of optimizing the lateral stiffness of a building while allowing an instantaneous update of the wind-induced structural design loads on the building. Specifically, an integration of the wind tunnel-based aerodynamic load determination procedure and the lateral drift design optimization technique is presented.

In this chapter, an aerodynamic wind tunnel load analysis routine based on the HFFB or SMPSS wind tunnel techniques for determining ESWLs of a tall building with 3D modes is firstly introduced. The importance of wind load updating during structural design synthesis process is emphasized and highlighted. The optimization technique developed is based on the Optimality Criteria (OC) approach, which has been widely used and shown to be particularly effective for optimal element sizing design optimization of tall buildings by Chan (2001, 2004). By employing the proposed integrated analysis and optimization framework, any considerable structural modification expressed in terms of the change in the natural frequency of a building, will trigger off the wind load updating process using the measured generalized load spectra extracted from existing aerodynamic wind load databases. Finally, the effectiveness of the proposed integrated design optimization method is demonstrated by two tall building examples with and without complex 3D mode shapes. Not only is the OC optimization technique capable of searching for the optimum distribution of element stiffness of a tall building, but also is it able to reduce the wind-induced structural loads on the building, leading to a further saving in the consumption of structural materials while satisfying the serviceability wind drift criteria.

## 4.2    Determination of equivalent static wind loads

The basic principle of aerodynamic wind tunnel techniques is to measure the external aerodynamic loads of a prototype building using a lightweight rigid model which is geometrically similar to the test building and is free of vibration in the simulated wind. Once the aerodynamic wind loads are measured, the wind-induced structural loads and responses of the prototype building can then be solved analytically based on the dynamic properties of the building. In general, the wind-induced dynamics of buildings can be more conveniently analysed in the frequency domain.

Using aerodynamic loading information measured in the wind tunnel, the response of a building with 3D coupled mode shapes can be calculated by modal analysis for combinations of the building's mass, stiffness, damping and the oncoming wind speed. As the aerodynamic wind force is a random process, the $j$-th modal displacement response spectra of the building in the generalized coordinate system can be computed based on random vibration theory in the frequency domain (Chen and Kareem, 2005a) as

$$S_{q_j}(f) = \left| H_j(f) \right|^2 S_{Q_{jj}}(f) \qquad (4.1)$$

where $S_{q_j}(f)$ is the modal response spectrum of displacement; $S_{Q_{jj}}(f)$ is the input mode-generalized wind load spectrum; $H$ is the mechanical admittance function, its amplitude can be expressed as

$$|H_j(f)|^2 = \frac{1}{m_j^2 (2\pi f_j)^4} \frac{1}{\left[1-(f/f_j)^2\right]^2 + 4\xi_j^2 (f/f_j)^2}$$

(4.2)

where $f_j$ is the $j$-th modal frequency of the building; $\xi_j$ is the $j$-th modal damping ratio; and $m_j$ is the $j$-th modal mass.

The $j$-th mode-generalized force can be expressed in terms of base bending moments and torque in time domain as

$$Q_j(t) = \eta_{jy} M_x(t) + \eta_{jx} M_y(t) + \eta_{j\theta} M_\theta(t)$$

(4.3)

where $\eta_s$ $(s=x,y,\theta)$ represents the mode shape correction that depends on the manner of mode shapes as well as the local wind pressure distribution over a building surface (Chen and Kareem, 2005a). Based on eq. (4.3), the corresponding power spectral density of $Q_j(t)$ are given as

$$S_{Q_{jj}}(f) = \sum_{s=x,y,\theta} \sum_{l=x,y,\theta} \eta_{js} \eta_{jl} S_{M_{sl}}(f)$$

(4.4)

where $S_{M_{sl}}(f)$ denotes the auto or cross power spectra density (XPSD) between the respective base moments $M_s(t)$ and $M_l(t)$. It was noted that cross-spectra in terms of XPSD can make significant contributions to the autospectra in the modal coordinate system.

In the situation that the aerodynamic wind loads are to be measured by the SMPSS technique, the $j$-th modal force can then be more accurately evaluated without the need for mode shape corrections as

$$Q_j(t) = \int_0^H [P_x(z,t)\Phi_{jx}(z) + P_y(z,t)\Phi_{jy}(z) + P_\theta(z,t)\Phi_{j\theta}(z)]dz$$

(4.5)

where $P_s(z,t)$ $(s=x,y,\theta)$ =the component wind loads per unit height at the elevation $z$ above the ground measured by the SMPSS technique; $\Phi_{js}(z)$ $(s=x,y,\theta)$ = the component of the $j$-th 3D mode shape. Once the modal force spectra are available, the mean square generalized displacement response can be calculated by integrating the area underneath the displacement response spectrum curve given in Eq. (4.1) as

74

$$\sigma_{q_j}^2 = \int_0^\infty S_{q_j}(f)df = \int_0^\infty |H(f)|^2 S_{Q_{jj}}(f)df \qquad (4.6)$$

Given the modal displacement for the $j$-th vibration mode, the mean square base moment or torque response can be calculated by combining the modal components using the complete quadratic combination (CQC) approach as follows

$$\sigma_{M_s}^2 = \sum_{j=1}^n \sum_{k=1}^n \sigma_{M_{js}} \sigma_{M_{ks}} r_{jk} = \sum_{j=1}^n \sum_{k=1}^n \Gamma_{js} \Gamma_{ks} \sigma_{q_j} \sigma_{q_k} r_{jk} \qquad (4.7)$$

in which the first $n$ fundamental modes are generally considered with the assumption that the contribution from the higher modes is negligible; $\sigma_{M_{js}}$ = the standard deviation value of the $j$-th modal component of the base bending moment or torque; $\Gamma_{js}, \Gamma_{ks}$ = the modal participation coefficients of the s-th component of the base moment; $\sigma_{q_{jr}}$ = the standard deviation value of the $j$-th modal displacement, $r_{jk}$ = the intermodal correlation coefficient for the $j$-th and $k$-th modal response. The detailed evaluation of the intermodal correlation coefficient $r_{jk}$ has been extensively discussed in Chapter 3, and can be referred to the literature of Xie et al (2003), Chen and Kareem (2005b) and Huang et al (2007).

In general, the mean square response can be approximated as the sum of the background component and the resonant component as follows (Davenport 1995)

$$\sigma_{M_s}^2 = \sigma_{M_{bs}}^2 + \sigma_{M_{rs}}^2 \qquad (4.8)$$

in which the first part of the right-hand side is referred to as the background component, which can be considered as quasi-static and can be directly quantified by integrating the area underneath the corresponding base moment spectrum curve as

$$\sigma_{M_{bs}}^2 \approx \int_0^\infty S_{M_s}(f)df \qquad (4.9)$$

The second part of the right-hand side of Eq. (4.8) represents the resonant dynamic amplification. For a lightly damped structural system, such as a tall building, the resonant component of the mean square base moment response can be obtained approximately using the modal superposition method as

$$\sigma_{M_{rs}}^2 = \sum_{j=1}^n \sum_{k=1}^n \sigma_{M_{jrs}} \sigma_{M_{krs}} r_{jk} = \sum_{j=1}^n \sum_{k=1}^n \Gamma_{js} \Gamma_{ks} \sigma_{q_{jr}} \sigma_{q_{kr}} r_{jk} \qquad (4.10)$$

75

where $\sigma_{M_{jrs}}$ = the standard deviation value of the $j$-th modal component of the resonant base bending moment or torque; $\sigma_{q_{jr}}$ = the standard deviation value of the resonant component of the $j$-th modal displacement, which can be simplified into the following algebraic euqaiton using the white noise assumption as

$$\sigma_{q_{jr}}^2 \approx S_{Q_{jj}}(f_j) \int_0^\infty |H_j(f)|^2 \, df = \frac{1}{m_j^2 (2\pi f_j)^4} \frac{\pi}{4\xi_j} f_j S_{Q_{jj}}(f_j) \qquad (4.11)$$

By substituting Eq. (4.11) into Eq. (4.10), the standard deviation resonant component $\sigma_{M_{rs}}$, of the base moments response $M_s$ can be explicitly expressed in terms of generalized force spectra as

$$\sigma_{M_{rs}}^2 = \sum_{j=1}^n \sum_{k=1}^n \frac{\Gamma_{js}\Gamma_{ks}r_{jk}}{64\pi^3 m_j m_k \left(f_j f_k\right)^2} \sqrt{\frac{f_j f_k}{\xi_j \xi_k}} S_{Q_{jj}}(f_j) S_{Q_{kk}}(f_k) \qquad (4.12)$$

Using the gust response factor approach, the peak base moment or torque response can be rewritten as (Kareem and Zhou 2003)

$$\hat{M}_s = \bar{M}_s + \sqrt{\hat{M}_{bs}^2 + \hat{M}_{rs}^2} \qquad (4.13)$$

where the background component $\hat{M}_{bs}$ and the resonant component $\hat{M}_{rs}$ can be calculated from Eqs. (4.9), (4.10), respectively as

$$\hat{M}_{bs} = g_b \sigma_{M_{bs}} \qquad (4.14)$$

$$\hat{M}_{rs} = g_r \sigma_{M_{rs}} = g_r \left( \sum_{j=1}^n \sum_{k=1}^n \frac{\Gamma_{js}\Gamma_{ks}r_{jk}}{64\pi^3 M_j M_k \left(f_j f_k\right)^2} \sqrt{\frac{f_j f_k}{\xi_j \xi_k}} S_{Q_{jj}}(f_j) S_{Q_{kk}}(f_k) \right)^{1/2} \qquad (4.15)$$

in which the background peak factor, $g_b$, can be approximated by the gust factor of the oncoming wind velocity, the value of which is usually taken to be about 3 to 4 (Zhou et al. 1999). For a Gaussian process, the resonant peak factor, $g_r$, can be given as (Davenport 1967)

$$g_r = \sqrt{2\ln v \cdot T} + \frac{0.577}{\sqrt{2\ln v \cdot T}} \qquad (4.16)$$

where $T$ is the observation time (usually 3600 s) for the given wind conditions under consideration; $v$ indicates the mean zero-crossing rate for the base moment component process, and could be fairly approximated by the first modal frequency value.

With the derived base moments and modal force spectra curves, the expected peak base moment or torque responses are calculated by Eq. (4.13). Based on the calculated peak base moments or torque, the wind-induced structural loads (or the so-called ESWLs) on the building can then be determined by distributing the peak base moments or torque to the floor levels over the building height (Holmes 2002; Chen and Kareem 2004). Similar to the base moments, the equivalent static wind loads expressed in terms of peak load $\hat{F}_s$ at each floor level, can also be written into a linear combination of the mean ($\bar{F}_s$), the background ($W_b \hat{F}_b$) and the resonant ($\sum_{j=1}^{n} W_{jrs} \hat{F}_{jrs}$) components as

$$\hat{F}_s = \bar{F}_s + W_b \hat{F}_{bs} + \sum_{j=1}^{n} W_{jrs} \hat{F}_{jrs}, \qquad (s = x, y, \theta) \qquad (4.17)$$

where $W_b = \dfrac{\sigma_{M_b}}{(\sigma_{M_b}^2 + \sigma_{M_r}^2)^{1/2}}$ , $W_{jrs} = \dfrac{\sum_{k=1}^{n} \sigma_{M_{jrs}} r_{jk}}{(\sigma_{M_b}^2 + \sigma_{M_r}^2)^{1/2}}$ ; $\hat{F}_{bs}$ is the $s$-th component peak background wind loads and $\hat{F}_{jrs}$ is the $s$-th component of the $j$-th modal peak resonant wind loads. Using Eq. (4.17), the ESWLs can be expressed respectively in the two translational and one torsional directions of a building in association with the default global coordinate system. In theory, the mean component of both the crosswind and torsional wind loads should be equal to zero. The alongwind mean force can be related to the approaching wind velocity profile and written as follows

$$\bar{F}(z) = \frac{1}{2} \rho \bar{U}_H^2 \left( \frac{z}{H} \right)^{2\alpha} BC_D \qquad (4.18)$$

where $\rho$=the air density; $\bar{U}_H$=the wind speed at the top of the building; $\alpha$=the power law exponent of wind profile; B=the width of the building; $C_D$= the drag force coefficient of the building. Due to the quasi-static nature of the background component of the wind loads, the distribution of the background component ESWLs to the floor levels over the building height can be assumed to follow the distribution of the mean alongwind loading profile given in Eq. (4.18) as (Zhou et al. 1999)

$$\hat{F}_{bx,by}(z) = \frac{\bar{F}(z)}{\displaystyle\int_0^H \bar{F}(z) z \, dz} \hat{M}_{bx,by} \qquad (4.19)$$

77

$$\hat{F}_{b\theta}(z) = \frac{\overline{F}_x(z)z}{\int_0^H \overline{F}_x(z)zdz}\hat{M}_{b\theta} \qquad (4.20)$$

where $\hat{F}_{bs}(z)$ $(s = x, y, \theta)$ is the distributed peak background wind load.

For a building having 3D coupled mode shapes, the distribution of the resonant component of the wind loads follows basically the mass-related modal inertial force distribution as shown in the following

$$\hat{F}_{jrs}(z) = g_r\sigma_{M_{jrs}}\mu_s(z) = g_r\Gamma_{js}\sigma_{q_r}\mu_s(z) \qquad (4.21)$$

where $\hat{F}_{jrs}(z)$ $(s=x,y,\theta)$ is the peak resonant wind load corresponding to the $j$-th mode per unit height; $\mu_s(z)$ $(s=x,y,\theta)$ is the inertial distribution factor, and is given as

$$\mu_s(z) = \begin{cases} \dfrac{m(z)\Phi_{jx,jy}(z)}{\int_0^H m(z)\Phi_{jx,jy}(z)zdz}, & s = x, y \\[4mm] \dfrac{I(z)\phi_{j\theta}(z)}{\int_0^H I(z)\phi_{j\theta}(z)dz}, & s = \theta \end{cases} \qquad (4.22)$$

where $m(z)$ is the mass per unit height; $I(z)$ is the rotational mass moment of inertia about the vertical axis per unit height.

## 4.3 Dependence of wind-induced loads on natural frequency

For multistory buildings, the ESWLs corresponding to the specific incident wind angle are generally expressed in terms of the alongwind, crosswind and torsional directions; and each directional ESWL consists of the mean, background and resonant components. Generally speaking, the mean and background components are the major parts of wind loads in the alongwind direction, since the mean wind loads quantified by Eq. (4.18) theoretically only exist in the alongwind direction. On the other hand, the resonant component may be the critical element of wind loads expressed in the crosswind and torsional directions. For wind-sensitive tall buildings, significant contributions of the resonant component to the total wind effects on buildings can be found in the crosswind as well as torsioanl directions. The crosswind loading effects is mainly due to wake excitation by vortex shedding as well as turbulence and buffeting by wind flow re-attachment on the

78

faces of the building. Several causes of the torsion can be identified, for example, uneven distribution of fluctuating wind pressures, building shape, interfering effects of nearby buildings, and dynamic characteristics of the structural system. Aerodynamically, the presence of turbulent flows of various scales and the existence of neighboring building structures may cause the uneven distribution of fluctuating wind pressures. The unbalance in aerodynamic force and uneven distribution of fluctuating wind pressures are the major aerodynamic sources of torsional wind loads on buildings. Even for a symmetric building, the instantaneous fluctuating torsional aerodynamic force may still exist.

Since the mean and the background wind loading components are mainly related to the approaching wind conditions and the aerodynamic shape of the building, they are very much independent of the building natural frequency (Davenport 1995). As a result, varying the building's natural frequency by modifying the structural stiffness of the building normally has a negligible effect on the mean and background values of the wind-induced structural forces. On the other hand, as given in Eq. (4.15), the resonant component of the wind loads is directly related to the natural frequency, $f_j$, of the building such that any changes in the natural frequency of the structure will subsequently lead to a direct impact on the wind-induced resonant effects on the building, particularly in the crosswind and torsional directions. For slender and compact tall buildings, the wind induced responses are mostly dominated by vortex shedding resonant effects in crosswind direction (Kareem 1985). For buildings with elongated plan forms, the resonant component of torsional wind loading also has a significant contribution to the total wind-induced response of such buildings (Boggs, et al., 2000). The torsional effects are further accentuated in the presence of eccentricities between the rigidity center, mass center and aerodynamic force center and for buildings with relatively long torsional periods of vibration. Accordingly, the ESWLs for those wind sensitive tall building structures are also generally found to be influenced more by the resonant effects in the crosswind and torsional directions.

Once the wind tunnel derived aerodynamic wind load spectra are available, they can be explicitly expressed in the form of an algebraic function of the modal frequency using regression analysis (Islam et al. 1992; Chan and Chiu 2006). Such algebraic functions can then be used to directly update the prediction of wind-induced structural loads on the building for any instantaneous change in the dynamic properties of the building during the design optimization synthesis. With the aid of piece-wise regression analysis, the power spectra density function of modal wind forces for a typical tall building can be inversely related to the modal frequency of the building and is explicitly expressed as an algebraic function in term of the modal frequency within the typical range of frequency for serviceability check, characterizing the descending part of the power spectra, as follows

$$S_{Q_{jj}}(f_j) \approx \beta_j f_j^{-\alpha_j} \qquad (4.23)$$

where $\alpha_j$ and $\beta_j$ are regression constants and normally $\alpha_j > 1$ and $\beta_j > 0$. Substituting Eq. (4.23) into Eq. (4.11), and subsequently into (4.21) gives the direct dependence of the resonant components of ESWLs to the modal frequency as

$$\hat{F}_{jrs}(z) \approx f_j^{-(\alpha_j+3)/2} \frac{g_r \Gamma_{js}}{8m_j \pi^{3/2}} \sqrt{\frac{\beta_j}{\xi_j}} \mu_s(z) \qquad (4.24)$$

For wind sensitive tall buildings where the value of the exponent $\alpha_j$ is normally greater than 0, the resonant ESWLs can be reduced by increasing modal frequency according to Eq. (4.24).

The resonant components of ESWLs would be combined together with the mean components as well as background components using Eq. (4.17) to obtain peak design wind load at each floor level for design purpose. For normal low-rise buildings, wind-induced structural loads are dominated by static mean components and quasi-static background components. The wind-induced resonant effects on low-rise buildings are small and negligible such that wind loads can be considered as constant static design loads. However, for dynamically sensitive tall buildings, wind-induced resonant effects become critical. In order to make an accurate prediction of the wind-induced structural loads on the building, it is necessary that the ESWLs be always updated whenever there exists a significant change in the structural properties of the building. In general, the resonant component of the ESWLs for a tall building can be reduced by increasing the modal frequencies of the building through structural optimization by efficiently distributing structural materials to improve the lateral and tensional stiffness of the building.


## 4.4 Design optimization


### 4.4.1 Formulation of lateral drift design problem

Consider a general tall building having $i_s = 1,2, ..., N_s$ steel frame elements, $i_c = 1,2, ..., N_c$ concrete frame elements and $i_w = 1,2, ..., N_w$ concrete shear wall elements. The design problem of seeking for the minimum material cost design of the building structure subject to $j = 1,2,..., N_g$ multiple lateral drift design constraints can be stated as:

Minimize structural materials cost

$$W(A_{i_s}, B_{i_c}, D_{i_c}, t_{i_w}) = \sum_{i_s=1}^{N_s} w_{i_s} A_{i_s} + \sum_{i_c=1}^{N_c} w_{i_c} B_{i_c} D_{i_c} + \sum_{i_w=1}^{N_w} w_{i_w} t_{i_w} \qquad (4.25)$$

Subject to

$$d_j \leq d_j^U \qquad (j = 1...N_g) \qquad\qquad (4.26)$$

$$A_{i_s}^L \leq A_{i_s} \leq A_{i_s}^U \qquad (i_s = 1...N_s) \qquad\qquad (4.27)$$

$$B_{i_c}^L \leq B_{i_c} \leq B_{i_c}^U \qquad (i_c = 1...N_c) \qquad\qquad (4.28)$$

$$D_{i_c}^L \leq D_{i_c} \leq D_{i_c}^U \qquad (i_c = 1...N_c) \qquad\qquad (4.29)$$

$$t_{i_w}^L \leq t_{i_w} \leq t_{i_w}^U \qquad (i_w = 1...N_w) \qquad\qquad (4.30)$$

Eq. (4.25) defines the minimum material cost design objective, in which $A_{i_s}$ is the cross-section area of steel element $i_s$, $B_{i_c}, D_{i_c}$ are the breadth and depth dimensions of rectangular concrete frame element $i_c$, and $t_{i_w}$ is the thickness concrete shear wall element $i_w$, respectively; and $w$ is the corresponding unit material cost. Eq. (4.26) defines the set of $j = 1,2,..., N_g$ interstory drift or top deflection constraints under the equivalent static wind load conditions where $d_j^U$ represents the predefined allowable interstory drift or overall top deflection limit. In general, the allowable drift ratio for buildings appears to be within the range of 1/750 to 1/250, with 1/400 being typical (Ad Hoc Committee, 1986). Eq. (4.27) to (4.30) define the element sizing constraints in which superscript $L$ denotes the lower size bound and superscript $U$ denotes the upper size bound of member $i$. It should be noted that the lateral drift response given in Eq. (4.26) for a general 3D building represents always the maximum resultant drift value at the most critical corner of the building (see Figure 4.3) since the ESWLs derived for a building even under unidirectional approaching wind condition are three dimensional in general.

81

To facilitate a numerical solution of the design optimization problem, the implicit drift constraints in Eq. (4.26) must be formulated explicitly in terms of the design variables $A_{i_s}, B_{i_c}, D_{i_c}, t_{i_w}$. Using the principle of virtual work, the collective set of lateral drift constraints can be expressed explicitly as

$$
d_j(A_{i_s}, B_{i_c}, D_{i_c}, t_{i_w}) = \sum_{i_s=1}^{N_s} \left( \frac{e_{i_s j}}{A_{i_s}} + e'_{i_s j} \right) + \sum_{i_c=1}^{N_c} \left( \frac{e_{0i_c j}}{B_{i_c} D_{i_c}} + \frac{e_{1i_c j}}{B_{i_c} D_{i_c}^3} + \frac{e_{2i_c j}}{B_{i_c}^3 D_{i_c}} \right)
$$

$$
+ \sum_{i_w=1}^{N_w} \left( \frac{e_{0i_w j}}{t_{i_w}} + \frac{e_{1i_w j}}{t_{i_w}^3} \right) \leq d_j^U \tag{4.31}
$$

where $e_{ij}$ and $e_{ij}'$ are the virtual strain energy coefficient and its correction factor of steel members respectively; $e_{0ij}$, $e_{1ij}$ and $e_{2ij}$ are the virtual strain energy coefficients of concrete members (Chan 2001). Once the finite element analysis is carried out for a given structural design under the ESWLs and virtual loading conditions, the internal element forces and moments are obtained and the element's virtual strain energy coefficients can then be readily calculated.

### 4.4.2 Optimality Criteria method

Upon establishing the explicit formulation of the drift constraints, the next task is to apply a suitable numerical technique for solving the optimal serviceability design problem. A rigorously derived Optimality Criteria (OC) method, which has been shown to be computationally efficient for large-scale structures is herein employed (Chan 2001, Chan 2004). In this OC approach, a set of optimality criteria for the optimal design is first derived and a recursive algorithm is then applied to indirectly solve for the optimal solution by satisfying the derived optimality criteria.

For simplification of discussion, the optimal drift design problem (i.e., Eqs. (4.25) to (4.30)) can be written into a generic form in terms of generalized member design variable $z_i$ representing all element sizing design variables (i.e., $A_{i_s}, B_{i_c}, D_{i_c}$ and $t_{i_w}$)

Minimize the structural material cost function

$$
W(z_i) \qquad (i = 1, 2, .., N_T) \tag{4.32}
$$

Subject to

$$
g_j(z_i) = \frac{d_j}{d_j^U} \leq 1 \qquad (j = 1, 2, .., N_j) \tag{4.33}
$$

$$z_i^L \leq z_i \leq z_i^U \quad (\ i = 1, 2, .., N_T) \tag{4.34}$$

To seek for numerical solution using the OC method, the constrained optimal design problem (i.e., Eqs. (4.32) to (4.34)) must be transformed into an unconstrained Lagrangian function which involves both the objective function and the set of explicitly expressed drift constraints of $g_j(z_i)$ associated with corresponding Lagrangian multipliers as

$$L\left(z_i, \lambda_j\right) = W\left(z_i\right) + \sum_{j=1}^{N_j} \lambda_j \left[ g_j(z_i) - 1 \right] \tag{4.35}$$

where $\lambda_j$ denotes the associated Lagrangian multiplier for the $j$-th design constraint. By differentiating the Lagragian function with respect to each sizing design variable and setting the derivatives to zero, the necessary stationary optimality conditions can be obtained as

$$\frac{\partial L}{\partial z_i} = 0 \Rightarrow \frac{\partial W}{\partial z_i} + \sum_{j=1}^{N_j} \lambda_j \frac{\partial g_j}{\partial z_i} = 0 \quad (\ i = 1, 2, .., N_T) \tag{4.36}$$

To avoid situations in which the stationary condition of Eq. (4.36) are satisfied and yet $\mathbf{z}^T = \left\{ z_1, z_2, \cdots, z_{N_T} \right\}$ is not a local minimum, the Lagrangian multipliers must be nonnegative, i.e.,

$$\lambda_j \geq 0 \quad (\ j = 1, 2, .., N_j) \tag{4.37}$$

Eqs. (4.36) and (4.37) are the KKT conditions for a relative minimum in a nonlinear programming problem defined by Eqs. (4.32) to (4.34) (Karush 1939; Kuhn and Tucker 1951).

The Eq. (4.36) could be written as a weighted gradient ratio

$$\frac{-\sum_{j=1}^{N_j} \lambda_j \dfrac{\partial g_j}{\partial z_i}}{\dfrac{\partial W}{\partial z_i}} = -\sum_{j=1}^{N_j} \lambda_j \left( \frac{-\dfrac{\partial g_j}{\partial z_i}}{\dfrac{\partial W}{\partial z_i}} \right) = 1 \tag{4.38}$$

The above equation is called the optimality criteria, which indicates that $\mathbf{x}^T$ reaches its optimum value when the weighted gradient ratio approaching to unity. If the drift constraints $g_j$ are formulated by the principle of virtual work as in Eq. (4.31) using the virtual strain energy coefficient, i.e., $e_{ij}$ and $e_{ij}'$, the optimality criteria could be interpreted as the weighted sum of virtual strain energy densities for each structural element must be equal to unity for an optimal solution.

The optimality criteria of Eq. (4.38) could be utilized in the following recursive relation to resize the active sizing variables (Chan 2001) as

$$z_i^{v+1} = z_i^v \cdot \left\{ 1 - \frac{1}{\eta} \left( \sum_{j=1}^{N_j} \lambda_j \frac{\partial g_j / \partial z_i}{\partial W / \partial z_i} + 1 \right) \right\}_v \qquad (i = 1, 2, .., N_T) \qquad (4.39)$$

where $v$ represents the current iteration number; and $\eta$ is a relaxation parameter as a step size parameter to control the convergence rate of the recursive process. A large value of $\eta$ represents a smaller step size and vice-versa. During the resizing iteration for $x_i$, any member reaching its size bounds is deemed as an inactive member, and its size is set at its corresponding size limit. The two partial derivatives with respect to design variables involved in the optimality criteria and the recursive relationship could be readily obtained, since the material cost function $W$ and the drift constraints $g_j$ have been explicitly expressed in terms of the design variables.

Before Eq. (4.39) can be applied to resize $z_i$, the Lagrangian multipliers $\lambda_j$ must first be determined. Considering the sensitivity of the $k$-th constraint with respect to changes in the sizing design variables, one can derive a set of $N_j$ simultaneous linear equations to solve for the $N_j$ unknown $\lambda_j$ (Chan 2001):

$$\sum_{j=1}^{N_j} \lambda_j^v \left( \sum_{i=1}^{N_i} \frac{z_i \frac{\partial g_k}{\partial z_i} \frac{\partial g_j}{\partial z_i}}{\frac{\partial W}{\partial z_i}} \right) = -\sum_{i=1}^{N_i} \left( z_i \frac{\partial g_k}{\partial z_i} \right)_v - \eta \left( g_k^{v+1} - g_k^v \right) \qquad (k = 1, 2, .., N_j) \qquad (4.40)$$

Having the current design variables $z_i^v$, the corresponding $\lambda_j^v$ values are readily determined by solving Eq. (4.40). Having the current values of $\lambda_j^v$, the new set of design variables $z_i^{v+1}$ can, in turn, be obtained from Eq. (4.39). Therefore, the recursive application of the simultaneous equations of Eq. (4.40) to find the $\lambda_j^v$ and the resizing formula of Eq. (4.39) to find the new solution of design variables constitutes the OC algorithm. By successively applying the OC algorithm until convergence occurs, the optimal design solution is then found.

### 4.4.3  Procedure of design optimization

The proposed integration of the wind-induced load analysis and the design optimization procedure is outlined step by step as follows:

1. Given the geometric shape of a building, determine the aerodynamic wind load spectra by wind tunnel tests or the appropriate use of existing wind load data derived from well established

aerodynamic load databases of the NatHaz Modeling Laboratory.

2. Develop the finite element model for the building and carry out an eigenvalue analysis to obtain the natural frequencies and mode shapes of the building.

3. Based on the current set of dynamic properties of the building, determine the floor-by-floor ESWLs for the building using the wind load analysis procedure given in Section 4.2.

4. Apply the derived ESWLs to the building and carry out a static structural analysis to estimate the equivalent static peak drift responses of the building.

5. Establish the explicit expression of the drift constraints Eq. (4.31) and formulate explicitly the optimal drift design problem.

6. Using the current set of element sizes $x_i$, solve the system of simultaneous linear equations Eq. (4.40) for the set of Lagrange multiplier $\lambda_j$.

7. Using the current value of $\lambda_j^{\nu}$ to resize the new set of design variables $x_i^{\nu+1}$ in Eq. (4.39).

8. Check the convergence of the recursive process: if all $z_i^{\nu+1} = z_i^{\nu}$ and $\lambda_j^{\nu+1} = \lambda_j^{\nu}$, then proceed to step 9; otherwise, update Eq. (4.39) for the current $z_i^{\nu+1}$, set $\nu = \nu +1$ and return to step 6.

9. Check the convergence of the structure weight: if the weight of the structure for 3 consecutive design cycles is within prescribed convergence criteria, then terminate the design process with the minimum material cost of the structure; otherwise, return to step 2, update the wind loads and redo structural analysis for the next optimization cycle.

Figure 4.4 shows a schematic flow chart for the proposed integrated stiffness optimization and wind load analysis technique for a tall building design.


## 4.5   Illustrative example


### 4.5.1   Example 4-1: A 45-story CAARC building

#### 45-story rectangular framed-tube steel framework

A 45-story, 10-bay by 15-bay rectangular tubular steel framework as shown in Figure 4.5 was used to test the proposed integrated aerodynamic load analysis and design optimization procedure. With a story height of 4 m and a bay width of 3 m, the 45-story steel framework has an overall height of 180m and a rectangular floor plan dimension of 30 m by 45 m. The building has the same geometric shape of the standard CAARC standard building, which has long been used for calibration purposes by different wind tunnels (Melbourne, 1980). For the wind drift design of the example framework, a 50-year return period of wind at a speed of 41 m/s at the top of the building with a damping ratio of

2% in an urban environment was considered. A typical interstorey drift limit of 1/400 was imposed at the three critical corner columns of the building under the combined equivalent static alongwind, crosswind and torsional wind loads derived from the aerodynamic load database of the NatHaz Modeling Laboratory.

A rigid floor diaphragm with a lumped swaying mass of $6.75\times10^5$ kg and a rotational mass moment of inertia of $1.645\times10^8$ kg·m$^2$ at the geometric centre of each floor was assumed. Members were to be designed using AISC standard steel sections as follows: W30 shapes for beams and W14 shapes for columns. For ease of construction, the columns on each vertical column line were grouped together to have a common section over three adjacent stories, while the corner beams of the exterior frame were grouped together to have the same section on each floor, as were the beams near the centre of each face of the building.

In this example, only the wind loading approaching in the X-direction of the building was considered. Due to vortex shedding effects, significant crosswind movement induced by the oncoming X-direction wind was found in the Y-direction of the building. The resultant interstory drift ratios at the 3 critical corner columns were limited to be within the allowable value of 1/400. Two initial design cases were used to test the effectiveness of the proposed integrated aerodynamic wind load analysis and optimal element sizing design technique.

Case A: The initial member sizes were established on the basis of a preliminary strength check. Along the height of the building, the columns and beams were divided into five different zones as shown in Table 4.1. The lateral deflection profile of the initial design is shown in Figure 4.9. Since the initial design was developed primarily based on element strength requirements, its lateral drift responses were found to violate the threshold drift ratio limit of 1/400. The initial natural frequencies were found to be 0.197 Hz for the 1$^{st}$ mode, 0.251 Hz for the 2$^{nd}$ mode and 0.422 Hz for the 3$^{rd}$ mode.

Case B: In order to establish a more reasonable initial structure for the optimal drift design, the initial member sizes were modified directly using the simple scaling method. Given the initial case A structure weight of 8022 tons with an initial top deflection of 0.608m, the required total steel tonnage of the structure for satisfying the top drift limit of 0.45m alone (i.e. H/400 = 180/400) can be calculated as:

$$8022 \times (0.608/0.45) = 10838 \text{ tons}$$

As given in Table 4.1 the initial member sizes were increased by approximately 35% to satisfy the overall top drift requirement. Accordingly, the initial natural frequencies were also increased to

86

0.214 Hz for the 1$^{st}$ mode, 0.273 Hz for the 2$^{nd}$ mode and 0.463 Hz for the 3$^{rd}$ mode.

Table 4.2 presents the breakdown of wind-induced structural loads for the 45-story steel framework. It is shown in the Table that the mean and background components together have summed up totally 76% of the alongwind force, while the resonant component contributes to the remaining 24% of the alongwind force. Since the majority of the alongwind force is independent of the structural property of the building, the alongwind force could hardly be adjusted by the means of element sizing optimization alone unless considerable changes in the aerodynamic shapes or configuration of the building are allowed. However, in the crosswind direction, the resonant loading is found to be the most dominating component that contributes to 82% of the total crosswind force. As the resonant loading is directly related to the modal frequency of the building, any considerable modification in the dynamic properties by changing as the building's stiffness, mass and mode shapes will therefore result in a more significant change in the value of the crosswind force on the building. It is also noted that the resonant component contributes only to a small portion of the total torsional wind load when compared with the crosswind load as shown in Table 4.2. Similar result has also been noted in a recent work by Chen and Kareem (2005a). Since the torsional mode usually represents a higher mode of vibration of a building, the normalized spectral density value corresponding to the higher mode of vibration due to torsion is generally quite small when compared to that of the crosswind sway mode of vibration, resulting in a smaller value of RMS resonant response. Nevertheless, it is worth to note that for buildings with elongated plan forms, the resonant component of torsional wind loading also has a significant contribution to the total wind-induced response of such buildings as shown in the second example.

Table 4.3 presents the initial wind-induced structural loads for the two initial cases of the building. The initial values of the alongwind base shear, the crosswind base shear and the base torque for the strength-based case A building were found to be larger than that of the stiffer case B building. Apparently, the increase in the stiffness from the strength-based case A design to the top-drift case B design has caused a relatively larger percentage reduction in the crosswind load (9.1%) when compared to that of the alongwind base shear (1.2%) and torsional moment (2.9%).

**Results and discussion**

Figures 4.6 (a) and (b) present the results of two design histories of the steel weight of the structure for cases A and B, respectively. One weight history was obtained using the integrated optimization method allowing for the use of the instantaneous wind load updating procedure, whilst another was obtained using the conventional optimization method without updating the wind load. It is found that the proposed integrated design optimization method incorporated with the wind load updating procedure could provide effectively the same final optimal design. On the other hand in the two cases that the conventional OC method was applied without updating the wind-induced structural loads, two different final designs were obtained due to the applications of two different sets of initial wind loads for cases A and B, respectively.

In case A, with a softer initial strength-based design structure resulting in a larger value of initial ESWLs, an increase of about 28.5% in the structure weight was needed to fulfill the interstory drift requirements if the conventional static optimization method without updating the wind load was employed. When using the instantaneous wind load updating procedure, the reduction of wind loads was taken into account so that a moderate increase of about 18.2% in the structure weight was found. As compared to the resulting steel material weight of 10310 tons obtained from the static optimization method without updating the wind load, the integrated optimization method has achieved a significant saving of about 10.3% in steel materials. In case B, 3.4% weight increase was found using the conventional OC method while the proposed integrated optimization technique was able to achieve about 0.7% saving in the material weight. It is noteworthy that the conventional optimization method assuming a fixed set of wind-tunnel derived loads resulted in a more conservative design since the reduction of the wind-induced loads due to the increase in the building stiffness was not taken into account in the design synthesis process. As shown in Figure 4.6, it is evident that a smooth and steady solution convergence was achieved when the applied wind load was kept unchanged throughout the design history. By comparison, a more bumpy fluctuating solution convergence was found especially during the first few design cycles for both cases A and B wherein the wind-induced structural loads were allowed to be updated after the optimization of each design cycle.

Figures 4.7 and 4.8 present graphically the variations of the alongwind, crosswind base shears and base torque throughout the design history of both cases of the building. At the end of each design cycle (including one formal structural analysis and one design optimization process), the alongwind, crosswind base shears and torsional base moment were updated. It is evident that the crosswind base shear was more significantly reduced as the stiffness of the structure was progressively improved by the OC technique. Increasing stiffness in the crosswind Y-direction by the optimization method reduced the crosswind base shear by 17.1% for case A and 8.8% for case B, respectively. As compared to the crosswind base shear, the alongwind base shear was found to only reduce slightly by 0.5% from 12051kN to 11996kN for case A. On the other hand for case B, the alongwind base shear was indeed slightly increased by 0.7% from 11906kN to 11996kN. The slight increase in the alongwind base shear can be explained by the fact that the original stiffness along the long X-direction of the building in case B was initially higher than necessary, and therefore was weaken by the OC optimization, leading to a saving in the structural material cost but a slight increase in the X-direction base shear. Although both the crosswind base shear and base torque are found somewhat fluctuating at the early stage of the design history, rapid and steady convergence to the final set of wind-induced loads is generally found after a few design cycles. Similar to the alongwind load, the torsion load is more dominated by the background component, the reduction in the torque load after the stiffness optimization is found relatively small for both cases A and B.

In both cases A and B with different initial starting designs, the OC technique was capable of producing almost the same optimal final design indicating that the OC method integrated with the wind load updating procedure is robust and somewhat insensitive to the value of the initial design. Since the weaker case A building was initially subjected to larger values of wind-induced structural loads as given in Table 4.3, a larger reduction in the wind loads for the case A building was found by the OC method.

Figures 4.9 and 4.10 present the initial and final lateral deflection profile and interstory drift ratio profile at the most critical corner column of the building, respectively. It is clearly shown that the initial case A design established solely based on strength check is found to violate considerably in both the lateral deflection and interstory drift ratio limits. By using a simple scaling method, the initial case B design was established to meet only the top deflection limit. Initially some violations in the interstory drift response of the case B design are found at the intermediate floor levels (i.e. between the $5^{th}$ and $35^{th}$ levels) of the building. After the optimization, no violation in the top deflection as well as the interstory drift responses is found in both cases A and B. While most story levels, particularly those intermediate levels, are found to have their interstory drift ratios to reach at or nearly close to the allowable limit of 1/400, the lateral deflection profiles of the optimized case A and B structures are found to be parallel closely to the allowable linear deflection profile as shown in Figures 4.9 and 4.10.

### 4.5.2   Example 4-2: A 40-story public housing building

#### 40-story public housing building

An optimization study of a 40-story residential building commissioned by the Hong Kong Housing Authority was carried out at the Hong Kong University of Science and technology. A three-dimensional computer model of the building and a structural layout plan are given in Figures 4.11 and 4.12, respectively. With an elongated width of 73 m, a narrow depth of 12 m and a total height of 122 m, the building has a critical aspect ratio (height/depth) of over 10.4. In view of its elongated and slender configuration, the building is anticipated to be wind sensitive and to exhibit significant swaying and twisting responses. The building is of reinforced concrete construction with coupled shear walls. As shown in Figures 4.11 and 4.12, multiple structural shear walls are coupled by lintel beams whenever possible to provide the total lateral and torsional resistance of the building. The effectiveness of the structural resisting system depends on many factors, such as the configuration of the structural form, the variable thickness of the shear walls and the variable dimensions of the lintel beams.

A wind tunnel test was carried out at the CLP Power Wind/Wave Tunnel Facility (WWTF) of the Hong Kong University of Science and Technology. One 1: 2000 scale topographical model incorporating relevant parts of the Hong Kong territory was firstly used to determine the representative approaching wind profiles for the building. Wind loads on the building were then measured by the HFFB technique using a 1:400 scale rigid model subjected to the specific wind profiles obtained from the topographical effect study. Wind tunnel measurements were taken for totally 36 incident wind angles at $10^{\circ}$ intervals for the full $360^{\circ}$ azimuth. Two critical incident wind directions corresponding to two perpendicular incident wind angles were identified by the wind tunnel test. One was the 0-degree wind perpendicular to the wide face acting in the short direction (i.e. along the Y-axis) of the building; another one was the 90-degree wind perpendicular to the narrow face acting in the long direction (i.e. along the X-axis). It was found that while the global maximum overturning moment about X-axis occurs at incident wind angle of 0 degree, the global maximum overturning moment about Y-axis and torsional moment about the vertical Z-axis occur at incident wind angle of 90 degree during the wind tunnel study. The power spectral density functions of modal forces for the building at two critical incident wind angles are shown in Figure 4.13. Provided the wind load spectra, the ESWLs on the building corresponding to these two most critical incident wind directions (0 degree and 90 degree wind) have been considered and updated during the design optimization process.

Table 4.4 presents a breakdown of the results of the maximum base shear forces and base torque for the initial building. It is noted that in Table 4.4 the maximum base shear force ($F_x$) and torsional moment ($M_{zz}$) are found under the 90 degree wind, and the maximum base shear force ($F_y$) is calculated under the 0 degree wind. Both maximum base shear forces ($F_x$ and $F_y$) occur in the alongwind direction corresponding to the two respective incident wind angles. As shown in Table 4.4 the mean and background components together have summed up to be slightly over 70% of the total alongwind loads for both base shears $F_x$ and $F_y$ respectively, while the resonant component contributes to the remaining 30% of each base shear. Since the base shear forces acting on the building are dominated by the quasi-static mean and background components, which are weakly dependent of the dynamic property of the building, only slight reduction in their values can be achieved by the means of stiffness design optimization. However, in the rotational direction about the vertical axis of the building, the dynamic resonant loading is found to be the dominating component that contributes to 72% of the total base torque. As the resonant component of wind loads is inversely proportional to the modal frequency of the building according to Eq. (4.24), any considerable modification in the dynamic stiffness of the building may led to a significant change in the torsional wind loading on the building.

In the preliminary design phase of the building before the wind tunnel test, wind loads on the building were calculated according to the Hong Kong Wind Code (HKCOP 2004). The alongwind loads in terms of base shear and base moment derived from the Hong Kong Wind Code are given in Table 4.5. The code-specified base shear $F_y$ is found to be slightly larger than the value predicted by wind tunnel test, while the base shear $F_x$ is substantially overestimated by the wind code. Unlike the Hong Kong Wind Code in which there is no specific guideline for predicting the wind-induced torsional load, the wind tunnel study gives a significant base torque of 201,340 kNm for the initial building, indicating an eccentricity of 8.7m equal to 11.8% of the width of the building.

The initial member sizes were established on the basis of a preliminary strength design. Once the finite element model having 2479 frame elements and 8523 shell panels was set up for the initial building, an eigenvalue analysis was then carried out to determine the dynamic properties of the building (i.e., the natural frequencies and mode shapes). The three-dimensional mode shapes of the building are given in Figure 4.14. The three fundamental coupled vibration modes have the first natural frequencies of 0.307 Hz (mainly torsional vibration), the second of 0.323 Hz (swaying primarily in the short direction of the building) and the third of 0.464Hz (swaying primarily in the long direction). The first vibration mode of the building is indeed a torsional mode, indicating significant dynamic torsional effects on the building.

In this stiffness optimization study, the major design variables are the thicknesses of variable shear walls. All variable shear wall elements for the design optimization task are highlighted as Wall Group 1~6 as shown in Figure 4.12. Due to some practical planning and constructability design considerations, the shear walls that are not highlighted in Figure 4.12 are kept unchanged during the optimization process. For this 40-story building, all shear walls are maintained to have uniform thickness, except Wall Group 3 in which three variations of wall thickness are allowed along the height of the building. Three vertical zones of wall thickness variations are allowed with zone one from ground floor to the 11$^{th}$ floor, zone two from the 12$^{th}$ floor to the 21$^{st}$ floor, and zone three from the $22^{nd}$ floor to the main roof of the building as illustrated in Fig. 4.11. Grade 45 concrete is used for the first 20 stories of the building and Grade 35 concrete for the upper 20stories. In the design optimization process, the top deflection limit of H/500, where H=114.1m is the height of the main roof from the ground, was imposed at the four top corners of the building.

**Results and discussion**

Figure 4.15 presents the design history of the total normalized cost of the structure. Although relatively stringent restrictions due to planning and constructability requirements have been imposed, the stiffness design optimization technique has achieved a 9.9% decrease of the total initial material cost of the building. A careful scrutiny on the optimized structure indicates that such a cost saving has been attained by more effectively stiffening the torsional resistance of the building through deepening the lintel beams at the two end walls (see Figure 4.12) and thickening the top parapet beam, which wraps around the building at the main roof level (see Figure 4.11). As a result of the torsional stiffness enhancement, the thickness of five out of the six variable shear walls, except wall group 5, can be reduced or kept constant as shown in Table 4.6. Consequently, the reduction in the wall thickness has also resulted in a net increase of 193 $m^2$ usable floor area.

As shown in Figure 4.15, a somewhat zigzag design history can be observed at the first few design cycles. At the beginning stage of the design history, as the structural stiffness of the building is improved by the OC optimization technique, there exists an increase in the frequencies of the structure and a corresponding reduction of the ESWLs, leading to a subsequent reduction in the lateral drift response. As a result of the reduced drift response, the structure tends to be weakened in the immediate following design cycle, thus causing a drop in the frequencies and subsequently an increase of the ESWLs, which, in turn, leads to an increased structural cost of the building. The end result gives a fluctuating zigzag design process, which converges to the final optimum design after 13 successive design cycles, producing a net 9.9% cost saving.

The results of the wind-induced base moments and torque of the building for the optimized structure are compared to that of the original structure as shown in Table 4.5. It is evident that only some minor reductions are found in the base shear forces ($F_x$ and $F_y$) and base overturning moments ($M_{xx}$ and $M_{yy}$), since their values are more dominated by the mean and background components of the wind loads. However, a more significant reduction in the base torque ($M_{zz}$) has been achieved for the optimized structure because of the fact that the value of the base torque is more dominated by the dynamic resonant component of the wind load. An increase in the torsional stiffness, by structural optimization, leading to an increase in the first modal frequency (torsional) of the building from 0.307 Hz to 0.318 Hz, has consequently resulted in the benefit of 8.7% reduction in the base torque.

With a significant wind-induced torsion applied on the residential building, the building is found to sway and twist with substantially larger deflection at the most distant corner of the building. As shown in Figure 4.16(a), the maximum top deflection of the initial structure at the most critical corner position is found to be 0.24m, giving a slight 5% violation in the top drift limit. It is worth noting that a much smaller deflection of 0.10m is found at the center of the top roof of the structure. An increase of 140% in the lateral deflection at the top corner of the initial building is due to torsional twisting induced by wind. After optimization, the maximum top drift ratio of the optimized structure is within the allowable drift ratio of 1/500 as shown in Figure 4.16(b). It is evident that the stiffness optimization technique is capable of achieving a more cost efficient design by redistributing the structural material to maximum the lateral and torsional stiffness of the building while satisfying the specified drift constraints.

## 4.6    Summary

This chapter presents the integrated aerodynamic wind load updating analysis and stiffness optimization technique for wind-induced drift design of tall buildings with 3D modes. Encouraging results have been found in the serviceability drift design optimization of a 45-story steel framework and a 40-story practical residential concrete building. The design optimization method integrated by wind engineering technique is able to produce the most cost efficient structural stiffness distribution of the building satisfying multiple lateral drift design constraints incorporating with torsional effects under multiple wind loading conditions. The integrated design optimization technique also is capable of achieving an additional benefit of wind load reduction by instantaneously updating wind-induced structural loads during the design synthesis process. While the crosswind base shear on the CAARC building has been significantly reduced by the integrated optimization technique, the wind-induced torsional loads on the public housing building with the asymmetric elongated plan form have been substantially reduced by the stiffness optimization method. It is expected that the optimal stiffness design methodology developed can be further extended to wind-induced dynamic serviceability design optimization (see Chapter 5) of mordern tall buildings considering both static drift and dynamic acceleration design criteria.

**Table 4.1    Initial member sizes of Cases A and B for the 45-story framework**

| Floor zone | Case A | | Case B | |
|---|---|---|---|---|
| | Column | Beam | Column | Beam |
| 37~45F | W14X159 | W30X211 | W14X257 | W30X292 |
| 28~36F | W14X257 | W30X261 | W14X370 | W30X326 |
| 19~27F | W14X370 | W30X292 | W14X455 | W30X357 |
| 10~18F | W14X500 | W30X326 | W14X500 | W30X391 |
| 1~9F | W14X550 | W30X357 | W14X550 | W30X433 |

**Table 4.2    Breakdown of wind-induced structural loads for the framework**

| Wind loads | Mean | Background | Resonant | Sum |
|---|---|---|---|---|
| Alongwind base shear (kN) | 6565 (54%) | 2538 (22%) | 2948 (24%) | 12051 (100%) |
| Crosswind base shear (kN) | 0 | 2223 (18%) | 10466 (82%) | 12689 (100%) |
| Torsional base torque (kNm) | 0 | 37106 (77%) | 11389 (23%) | 48495 (100%) |

**Table 4.3    Initial wind-induced structural loads for the 45-story framework**

| Wind loads | Alongwind base shear (kN) | Crosswind base shear (kN) | Torsional base torque (kNm) |
|---|---|---|---|
| Case A | 12051 | 12689 | 48495 |
| Case B | 11906 | 11533 | 47068 |
| Percentage of decrease | 1.2% | 9.1% | 2.9% |

### Table 4.4    Breakdown of maximum wind loads for the initial building

| Wind loads | Mean | Background | Resonant | Sum |
|---|---|---|---|---|
| Base shear $F_x$ (kN) | 1,075 (33%) | 1,221 (37%) | 962 (30%) | 3,258 (100%) |
| Base shear $F_y$ (kN) | 9,179 (40%) | 7,220 (31%) | 6,644 (29%) | 23,043 (100%) |
| Torsional base torque $M_{zz}$ (kNm) | 30,680 (15%) | 26,660 (13%) | 143,990 (72%) | 201,340 (100%) |

### Table 4.5    Wind-induced base shears and base moments of the 40-story building

| Wind loads | Hong Kong wind code | Wind tunnel results before optimization | Wind tunnel results after optimization | Wind load reduction |
|---|---|---|---|---|
| Base shear $F_x$ (kN) | 6,570 | 3,258 | 3,223 | -1.1% |
| Base shear $F_y$ (kN) | 24,170 | 23,040 | 22,690 | -1.5% |
| Base moment $M_{xx}$ (kNm) | 1530,500 | 1723,700 | 1702,800 | -1.2% |
| Base moment $M_{yy}$ (kNm) | 441,070 | 241,130 | 239,390 | -0.7% |
| Torsional base torque $M_{zz}$ (kNm) | 0 | 201,340 | 183,740 | -8.7% |

### Table 4.6    Original and optimized thickness of variable shear walls

| Wind Group | 1 | 2 | 3 | 4 | 5 | 6 |
|---|---|---|---|---|---|---|
| Original thickness (mm) | 400 | 525 | 525 | 300 | 300 | 525 |
| Optimized thickness (mm) | 400 | 300 | 350/250/200[*] | 200 | 425 | 300 |

[*]The three values of optimized thickness are corresponding to Zone1 to 3, respectively.

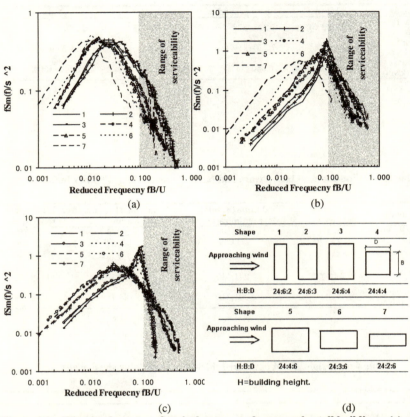

**Figure 4.1    Wind load spectra of typical square and rectangular tall buildings: (a) alongwind base moment spectra; (b) crosswind base moment spectra; (c) base torque spectra; (d) building cross-sections**

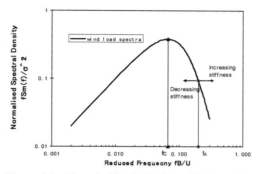

**Figure 4.2    The wind loads spectra with f$_A$ and f$_C$**

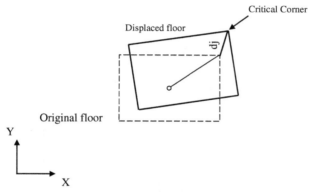

**Figure 4.3    Resultant drift at *j*-th story d$_j$**

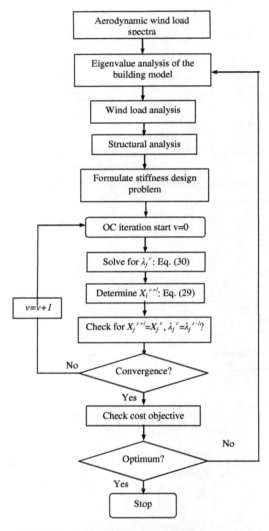

**Figure 4.4    Flow chart of integrated design optimization process**

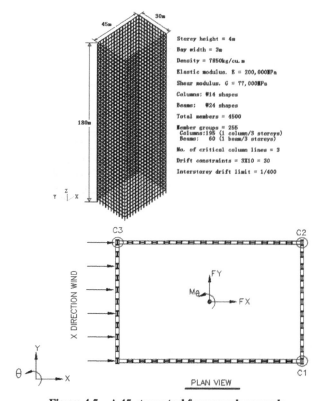

Storey height = 4m

Bay width = 3m

Density = 7850kg/cu.m

Elastic modulus, E = 200,000MPa

Shear modulus, G = 77,000MPa

Columns: W14 shapes

Beams:   W24 shapes

Total members = 4500

Member groups = 255
  Columns:195 (1 column/3 storeys)
  Beams:   60 (1 beam/3 storeys)

No. of critical column lines = 3

Drift constraints = 3X10 = 30

Interstorey drift limit = 1/400

PLAN VIEW

**Figure 4.5    A 45-story steel framework example**

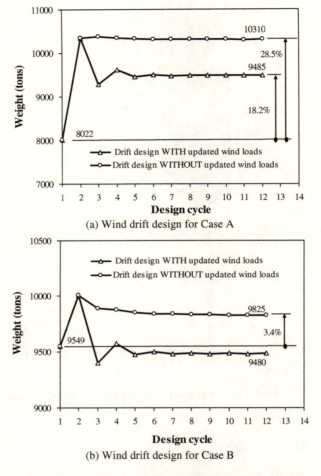

(a) Wind drift design for Case A

(b) Wind drift design for Case B

**Figure 4.6    History of the structure weight for the 45-story framework**

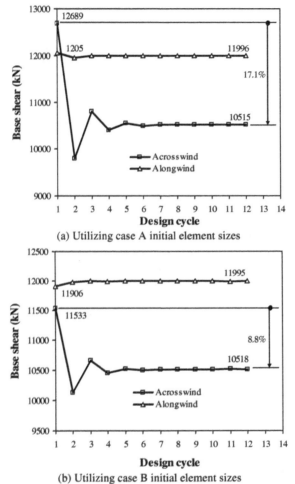

(a) Utilizing case A initial element sizes

(b) Utilizing case B initial element sizes

**Figure 4.7   History of the alongwind and crosswind base shear for the 45-story framework**

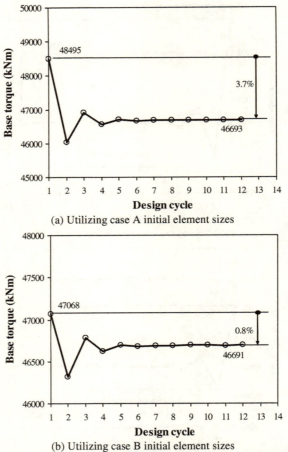

(a) Utilizing case A initial element sizes

(b) Utilizing case B initial element sizes

Figure 4.8    History of the base torque for the 45-story framework

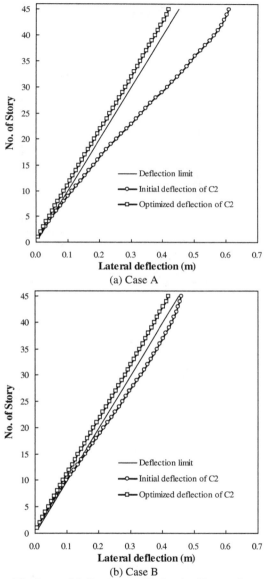

(a) Case A

(b) Case B

**Figure 4.9    Lateral deflection profile for the 45-story framework**

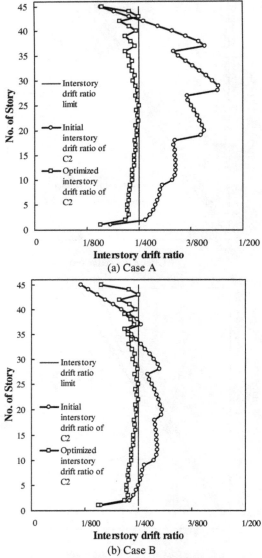

(a) Case A

(b) Case B

**Figure 4.10 Interstory drift ratio profile for the 45-story framework**

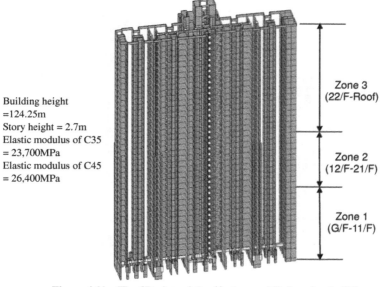

Building height
=124.25m
Story height = 2.7m
Elastic modulus of C35
= 23,700MPa
Elastic modulus of C45
= 26,400MPa

Zone 3
(22/F-Roof)

Zone 2
(12/F-21/F)

Zone 1
(G/F-11/F)

**Figure 4.11    The 3D view of the 40-story public housing building**

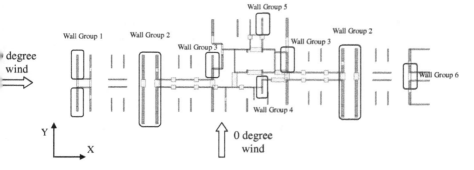

**Figure 4.12    Typical floor layout plan of with variable shear wall elements of the 40-story
public housing building**

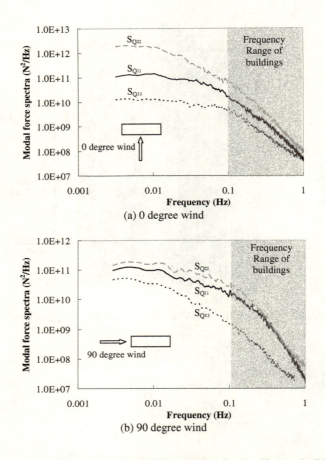

(a) 0 degree wind

(b) 90 degree wind

**Figure 4.13   Wind-induced modal force spectra for the 40-story building**

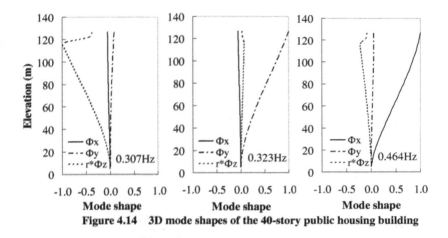

**Figure 4.14    3D mode shapes of the 40-story public housing building**

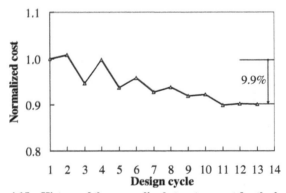

**Figure 4.15    History of the normalized structure cost for the building**

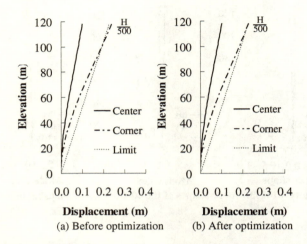

(a) Before optimization    (b) After optimization

**Figure 4.16    Lateral deflection profiles at the center and corner of the building**

# CHAPTER 5    Wind-induced Serviceability Design Optimization of Tall Buildings

## 5.1    Introduction

The earthquakes and strong winds are the two major hazard-related load conditions threatening to modern tall buildings. With the accumulated evidence of huge earthquake related losses found in the buildings all over the world (FEMA 2000a), leading structural engineers have promoted the development and application of performance-based seismic design (PBSD) concepts. The basic concept of PBSD is to provide engineers with the capability to design buildings that have a predictable and reliable performance in earthquake. While the performance-based seismic design (PBSD) is becoming well accepted in professional practice for the design of buildings located in a seismic-prone zone, the performance-based wind-resistant design (PBWD) is also emerging as a promising design methodology to improve the current practice in the tall building design against strong wind (e.g., typhoon).

Recent trends towards developing increasingly taller and irregularly-shaped buildings have led to slender complex structures that are highly sensitive and susceptible to wind-induced vibration. When today's tall buildings continue to increase in height, the occupant comfort or habitability evaluation of tall buildings becomes a more critical serviceability consideration in the structural design process. In the design of such new generation of tall buildings, structural engineers are facing the challenge of striving for the most efficient and economical design solution while ensuring that the final design must be serviceable for its intended function in addition to be safe over its design life-time. Therefore, there is a need to develop an integrated wind-induced dynamic analysis and performance-based design optimization technique for minimizing the structural cost of tall buildings subject to static drift and dynamic acceleration performance design criteria.

However, wind-induced performance levels and performance design objectives of tall buildings have not been well defined compared with the performance-based seismic engineering. An attempt to establish a compatible set of wind design criteria for multi-hazard performance-based design of high-rise buildings was made by Chock et al. (1998). The proposed design objectives were established to be "risk-consistent" with the framework for performance-based seismic engineering. Jain et al. (2001) proposed a probability based methodology used to determine site-specific performance-based design wind speeds for use in wind tunnel tests and building designs. It has demonstrated that using such site-specific extreme wind loads could often lead to more cost efficient designs. Generally, annual maximum wind speed (design wind seed) corresponding to 50-year (or any other time interval) return period of wind could be estimated by statistical analysis to best fit observed or simulated wind data into the cumulative distribution function of Type I asymptotic extreme value distribution (Zuranski and Jaspinska 1996; Minciarelli et al. 2001; Kasperski 2007).

The two major performance parameters for wind-induced serviceability design of tall buildings are the lateral displacement and the excessive acceleration. The extreme wind loading conditions of 50-year return period are generally used for checking the serviceability lateral drift criteria and the ultimate strength limit states for safety requirements. The checking procedure for motion perception can be performed by comparing the magnitude of wind-induced vibration and the acceptability threshold of motion (or the so-called, occupant comfort criteria). It has been widely accepted that the perception of wind-induced motion is closely related to the acceleration response of buildings (Kwok et al. 2007). Both peak acceleration and standard deviation acceleration under extreme wind conditions of 10-year or 5-year return period are commonly used to represent building motion (Burton et al. 2007). Frequency dependent motion perception criteria in terms of standard deviation acceleration have been established in the ISO Standard 6897 (ISO 1984) for low frequency motion due to sustained or regularly occurring wind events. The peak acceleration criteria associated with the extreme wind events have also been established in a number of modern design codes to preclude discomfort of occupants due to fear for safety in extreme windstorm or typhoon events. For example, the peak acceleration criteria of 15 milli-g for residential buildings and 25 milli-g for commercial buildings under 10-year return period of wind conditions are recommended in the National Building Code of Canada (NBCC 1995), the Chinese code (JGJ 3-2002), and the Hong Kong Codes of Practice (HKCOP 2004, HKCOP 2005). The Architectural Institute of Japan Guidelines for the Evaluation of Habitability (AIJ-GEH) (2004) offers frequency dependent peak acceleration criteria for evaluating wind-induced building vibration based on motion simulator investigations. The evaluation curve established based on experimental investigation by motion simulations specifies frequency dependent peak accelerations of one year recurrence of wind, which is believed to be more closely related to more frequent occupant comfort conditions.

The structural design task for a large-scale tall building satisfying multiple static and dynamic performance design criteria is a rather difficult and laborious task, which usually would be performed based on trial and error process by structural engineers. Optimization methods have long been recognized as a more rational approach to automatically seek the most cost efficient design subject to multiple design constraints. Performance-based design methodology allows a new perspective to formulate design optimization problems by taking into account either load uncertainties or engineering system uncertainties in the optimal performance-based design problem formulation, leading to the field of performance-based design optimization (PBDO). Ganzerli et al. (2000) addressed the optimal performance-based design of seismic structures, in which performance-based design concepts and pushover analysis were incorporated in the design of reinforced concrete structures using structural optimization. Foley (2002) presented a comprehensive review on the current state of research and development in the field of the optimal performance-based engineering. The need for more research effort in developing PBDO procedures was highlighted. The earthquake-induced drift performance optimization technique for reinforced concrete buildings was developed by Chan and Zou (2004) based on the Optimality Criteria (OC) method. Xu et al. (2006) presented a multicriteria optimization method for the performance-based seismic design of steel building frameworks under equivalent static seismic loading. The overall objective for the design of a building framework in that study is to have minimum structural weight and uniform plastic ductility demand while meeting displacement and strength constraints corresponding to various performance levels.

Recently, an effective design optimization approach has been developed for stiffness design optimization of symmetric tall buildings with uncoupled mode shapes subject to serviceability design criteria (Chan and Chui 2006; Chan et al. 2007). Chan and Chui (2006) presented an occupant comfort serviceability design optimization technique for symmetric tall steel buildings subject to the frequency dependent standard deviation acceleration criteria of ISO-6897. Chan et al. (2007) developed an integrated optimal design framework that couples together an aerodynamic wind load updating process and a drift design optimization algorithm for symmetric tall building structures. Although these research studies represent a major advance in the use of structural optimization techniques for wind-induced serviceability designs of tall buildings, it is necessary to extend the stiffness design optimization technique for general asymmetric tall buildings subject to multiple wind induced drift and acceleration performance design criteria.

In this chapter, an automated stiffness design optimization technique of tall buildings subject to the lateral static drift and dynamic acceleration performance design constraints is presented. The lateral static drift and dynamic peak and standard deviation acceleration responses of a tall building are analyzed under various levels of return period of wind loading corresponding to different wind-induced performance levels. The optimal stiffness design problem of a building subject to drift and both peak and standard deviation acceleration constraints is explicitly formulated in terms of element sizing design variables. A rigorously derived Optimality Criteria method, which has

111

been shown to be particularly suitable for large scale element sizing optimization problems, is employed to solve the optimal stiffness design optimization problem. Finally, a full-scale 60-story asymmetric building structure with 3D mode shapes is used to demonstrate the applicability and effectiveness of the developed automatic stiffness design optimization technique.

## 5.2 Probabilistic analysis of performance-based design wind speed

### 5.2.1 Estimation of design wind speed

For structural design purpose, extreme wind speed rather than daily wind speed is concerned. Therefore it seems to be reasonable to fit observed extreme wind speed data to the Type I Gumbel extreme value distribution. Denote the annual largest wind speed as a random variable $V$, which could be modeled by the Type I Gumbel extreme value distribution as

$$F_V(v) = P(V \le v) = 1 - p = \exp\left\{-\exp\left[-\frac{v-u}{\beta}\right]\right\} \qquad (5.1)$$

where $u$ is modal wind speed; $\beta$ is the dispersion or a scale wind speed; $p$ is the probability of the design wind speed $V$ being exceeded by a chosen value of $v$. Given a sample of $V$ with size $N$, the modal wind speed $u$ is defined as the particular value of $V$ such that the expected number of sample values larger than $u$ is one. The distribution parameters $u$ and $\beta$ could be estimated by the probability paper method (Ang and Tang 1975). Upon taking logarithms twice, Eq. (5.1) becomes

$$v = u + \beta\left\{-\ln\left[-\ln(1-p)\right]\right\} \qquad (5.2)$$

Hence the reciprocal of the slope of a graph of $-\ln\left[-\ln(1-p)\right]$ against $v$ is the estimation of $\beta$. The estimation of modal wind speed can also be obtained from a zero intercept on this graph. For a design wind speed of $V_R$ corresponding to a return period of $R$ years, the probability $p$ of the design wind speed $V_R$ being exceeded in one year is $p=1/R$, then the design wind speed $V_R$ could be estimated by

$$\tilde{V}_R = u + \beta\left\{-\ln\left[-\ln\left(1-\frac{1}{R}\right)\right]\right\} \qquad (5.3)$$

### 5.2.2 Uncertainties in the estimation of design wind speeds

Denoting the mean and standard deviation of the extreme wind speed by $E(V)$ and $D(V)$, respectively. From the knowledge of Type I extreme value distribution, the mean value and the standard deviation of the extreme wind speed can be calculated form the distribution parameters, the modal wind speed ($u$) and the dispersion of wind speed ($\beta$) as follows

$$E(U) = u + \gamma\beta \tag{5.4}$$

$$D(U) = \frac{\pi}{\sqrt{6}}\beta \tag{5.5}$$

where $\gamma$=0.5772 (the Euler constant). It is worth noting that $E(V)$ and $D(V)$ are assumed as the exact distribution characteristics of the whole population of extreme wind speeds regardless of a particular sample of extreme wind speed data. From Eq. (5.3), the design wind speed can be rewritten in term of $E(V)$ and $D(V)$ based on Eq. (5.4) and (5.5) as

$$V_R = E(V) + a(R)D(V) \tag{5.6}$$

where $a(R) = \sqrt{6}\left\{-\ln\left[-\ln\left(1-1/R\right)\right]\right\}/\pi - \sqrt{6}\gamma/\pi$.

Given a record $\{v_1, v_2, \cdots, v_N\}$ of the extreme wind speeds, the sample mean $\tilde{E}(V)$ and the sample standard deviation $\tilde{D}(V)$ of the extreme wind speed can be estimated as follows

$$\tilde{E}(V) = \frac{1}{n}\sum_{i=1}^{n} v_i \tag{5.7}$$

$$\tilde{D}(V) = \sqrt{\frac{1}{n}\sum_{i=1}^{n}\left[v_i - \tilde{E}(V)\right]^2} \tag{5.8}$$

The uncertainty in the estimation of design wind speeds can be evaluated by quantifying the mean and variance of the estimator of design wind speeds, $\tilde{V}_R$, which is obtained using the sample mean and the sample standard deviation instead of the exact mean and the exact standard deviation in Eq. (5.6) as follows

$$\tilde{V}_R = \tilde{E}(V) + a(R)\tilde{D}(V) \tag{5.9}$$

The value of $\tilde{V}_R$ is uncertain because the sample mean and sample standard deviation themselves are random variables whose variance depends on the sample size $N$. Based on the first and second moment characterization of the sample mean and sample standard deviation, the mean and variance of $\tilde{V}_R$ can be derived from Eq. (5.6) (Cramer 1946; Minciarelli et al. 2001) as follows

$$E\left(\tilde{V}_R\right) \approx E(V) + a(R)D(V) \tag{5.10}$$

$$D\left(\tilde{V}_R\right) \approx \sqrt{\frac{D(V)}{n}}\left[1 + a^2(R)\sqrt{\frac{\gamma_2 D^2(V) - 1}{4N}}\right] \tag{5.11}$$

where $\gamma_2$ denoting the kurtosis of the design wind speed distribution is defined as the fourth normalized moment to measure the degree of flattening of the curve of a probability density function near its center. The kurtosis for Type I extreme value distribution is given by Gumbel (1958) as $\gamma_2 = \left(3\pi^4\beta^4\right)/\left[20D^4(X)\right]$. Then the uncertainty in the estimation of design wind speeds could be quantified by the coefficient of variation (COV), i.e., $\mathrm{COV}\left(\tilde{V}_R\right) = D\left(\tilde{V}_R\right)/E\left(\tilde{V}_R\right)$. The uncertainty of design wind speeds in terms of the coefficient of variation is very useful for studying the propagation of the uncertainties of the parameters over the structural response as well as reliability-based performance assessment of wind-sensitive structures (Solari 1997).

## 5.3   Dynamic response analysis of wind-induced motion

Since wind excitation could be regarded as a stationary and ergodic random process, it is more convenient to conduct wind-induced response analysis in the frequency domain, as described in Chapter 3. Based on random vibration theory, the power spectral density (PSD) matrix of the displacement response vector can be calculated in the frequency domain using the full spectral approach.

Given the $s$ component of $n$ number of 3D mode shapes of the building system at elevation $z$, i.e., $\{\phi_s(z)\}^T = \{\phi_{1s}(z), \phi_{2s}(z), ..., \phi_{ns}(z)\}$, the mean square value of the $s$-th component acceleration response of the building system at elevation $z$ can be written in the form of modal combination as

$$\sigma_{\ddot{s}}^2 = \sum_{j=1}^{n}\phi_{js}^2(z)\sigma_{\ddot{q}_{jj}}^2 + \sum_{j=1}^{n}\sum_{\substack{k=1\\j\neq k}}^{n}\phi_{js}(z)\phi_{ks}(z)r_{jk}\sigma_{\ddot{q}_{jj}}\sigma_{\ddot{q}_{kk}}, \qquad (s = x, y, \theta) \tag{5.12}$$

where $\sigma_{\ddot{q}_{jj}}^2$ representing the variance or mean square value of the $j$-th modal acceleration response can be given as (Tallin and Ellingwood, 1985)

$$\sigma_{\ddot{q}_{jj}}^2 \approx \frac{\pi f_j}{4 \xi_j m_j^2} S_{Q_{jj}}(f_j) \tag{5.13}$$

where $m_j$ = the $j$-th modal mass.

In the case of a symmetrical building having uncoupled unidirectional mode shapes $\phi_j(z)$ in each principle direction, the building response in each respective direction can be considered separately in a mode-by-mode manner without the consideration of cross correlations of modal responses. Then the $s$-th component acceleration response of such a symmetrical building can be reduced from Eq. (5.12) as

$$\sigma_{\ddot{s}} = \phi_j(z)\sigma_{\ddot{q}_{jj}}, \qquad (s = x, y, \theta) \tag{5.14}$$

The total resultant acceleration response incorporating both swaying and torsional effects is an appropriate measure of the magnitude of building vibration located at a distance from the reference center of the top floor. For instance, the total translational component acceleration at the most distant corner ($R_x$, $R_y$) from the reference center of a building can be given as

$$a_{cx} = \ddot{x} - R_y \ddot{\theta}, \quad a_{cy} = \ddot{y} + R_x \ddot{\theta} \tag{5.15}$$

where $\ddot{x}, \ddot{y}$ = the two perpendicular translational acceleration components and $\ddot{\theta}$ = the torsional acceleration component at the reference center of the building. The resultant acceleration response acceleration $a_r$ at the corner point can be calculated as follows

$$a_r^2 = \ddot{x}^2 + \ddot{y}^2 + (R_x^2 + R_y^2)\ddot{\theta}^2 - 2R_y \ddot{x}\ddot{\theta} + 2R_x \ddot{y}\ddot{\theta} \tag{5.16}$$

By taking mathematical expectation operation $E(.)$ on both sides of the above equation, the mean square resultant acceleration response can be written as

$$E\left(a_r^2\right) = E\left(\ddot{x}^2\right) + E\left(\ddot{y}^2\right) + (R_x^2 + R_y^2)E\left(\ddot{\theta}^2\right) - 2R_y E\left(\ddot{x}\ddot{\theta}\right) + 2R_x E\left(\ddot{y}\ddot{\theta}\right) \tag{5.17}$$

Since the acceleration response components ($\ddot{x}, \ddot{y}, \ddot{\theta}$) are reasonably assumed to be zero mean Gaussian processes, their mean square values are equal to the corresponding variances. Therefore, the mean square resultant acceleration can be rewritten as

$$E\left(a_r^2\right) = \sigma_{\ddot{x}}^2 + \sigma_{\ddot{y}}^2 + (R_x^2 + R_y^2)\sigma_{\ddot{\theta}}^2 - 2R_y \sigma_{\ddot{x}\ddot{\theta}}^2 + 2R_x \sigma_{\ddot{y}\ddot{\theta}}^2 \tag{5.18}$$

where $\sigma_{\ddot{x}}^2, \sigma_{\ddot{y}}^2$ = the variance of the two translational acceleration components and $\sigma_{\ddot{\theta}}^2$ = the torsional acceleration component, all of which can be calculated using Eq. (5.12); $\sigma_{\ddot{x}\ddot{\theta}}^2$=covariance of $\ddot{x}$ and $\ddot{\theta}$ and $\sigma_{\ddot{y}\ddot{\theta}}^2$ = covariance of $\ddot{y}$ and $\ddot{\theta}$, which can be obtained by integrating off-diagonal PSD terms in the (PSD) matrix of the displacement response vector.

When the correlations between any two component motions in the $x$, $y$ and $\theta$ directions are small, the covariance terms $R_y\sigma_{\ddot{x}\ddot{\theta}}^2$ and $R_x\sigma_{\ddot{y}\ddot{\theta}}^2$ are small as compared with the other three variance terms such that the RMS resultant acceleration may be simplified to the following form as

$$\sqrt{E\left(a_r^2\right)} \approx \sqrt{\sigma_{\ddot{x}}^2+\sigma_{\ddot{y}}^2+(R_x^2+R_y^2)\sigma_{\ddot{\theta}}^2} \tag{5.19}$$

The mean value of resultant acceleration could be written as the following expression

$$E\left(a_r\right)= E\left[\sqrt{\ddot{x}^2+\ddot{y}^2+(R_x^2+R_y^2)\ddot{\theta}^2-2R_y\ddot{x}\ddot{\theta}+2R_x\ddot{y}\ddot{\theta}}\right] \tag{5.20}$$

which obviously have a value greater than zero. The standard deviation value of the resultant acceleration then could be expressed in terms of $E\left(a_r^2\right)$ and $E\left(a_r\right)$ as

$$\sigma_{a_r} = \sqrt{E\left(a_r^2\right)-\left[E(a_r)\right]^2} \tag{5.21}$$

The maximum peak resultant acceleration response over a given time duration $\tau$, is an important index for assessing occupant comfort performance of a tall building. The expected value of the largest peak resultant response at the top corner of the building, can be defined by

$$\hat{a}_{r,\tau} = E\left(a_r\right)+g_{a_r}\,\sigma_{a_r} \tag{5.22}$$

where $g_{a_r}$ = the peak factor for the resultant acceleration response process. However, the nonlinear transform of several Gaussian processes would result in a non-Gaussian process, e.g., the resultant acceleration process. The simple formulation of peak factor for a non-Gaussian process is generally not available (Gurley et al. 1997). In Chapter 6, a statistical approach would have been employed to estimate peak factors based on several data samples of any stationary response process. For a Gaussian process, Davenport gave a simple algebra formula to estimate peak factors as (Davenport 1964) as

$$g_f = \sqrt{2\ln\nu\tau}+0.5772/\sqrt{2\ln\nu\tau} \tag{5.23}$$

where $\nu$ denoting the mean zero-crossing rate of the Gaussian process can be approximated as the modal frequency of a building, and $\tau$ representing the observation time duration for assessing acceleration response may be normally taken as 600s.

As a conservative engineering approach (Isyumov et al. 1992), the expected maximum peak resultant response could be approximated by combining the two peak values of translational component acceleration at the corner. From Eq. (5.15), two peak values of translational component acceleration could be expressed in terms of the Davenport peak factor $g$, respectively

$$\hat{a}_{cx,\tau} \approx g_f \left( \sigma_{\dot{x}}^2 + R_y^2 \sigma_{\dot{\theta}}^2 \right)^{1/2} \tag{5.24}$$

$$\hat{a}_{cy,\tau} \approx g_f \left( \sigma_{\dot{y}}^2 + R_x^2 \sigma_{\dot{\theta}}^2 \right)^{1/2} \tag{5.25}$$

Then the expected maximum peak resultant acceleration response is given as

$$\hat{a}_{r,\tau} \approx \sqrt{\hat{a}_{cx,\tau}^2 + \hat{a}_{cy,\tau}^2}$$
$$= \sqrt{g_f \left( \sigma_{\dot{x}}^2 + R_y^2 \sigma_{\dot{\theta}}^2 \right) + g_f \left( \sigma_{\dot{y}}^2 + R_x^2 \sigma_{\dot{\theta}}^2 \right)} = g_f \sqrt{E\left(a_r^2\right)} \tag{5.26}$$

As a result, the expected maximum peak resultant acceleration response could be simply approximated by the multiplication of the RMS value $\sqrt{E\left(a_r^2\right)}$ and the Davenport peak factor $g_f$.

The above direct combination equation in Eq. (5.26) has implicitly assumed a full correlation among the directional maximum acceleration components. Such an assumption of the coincident occurrence of the maximum translational acceleration components generally gives an upper bound estimate on the maximum resultant response, thus resulting in unduly conservative prediction of the peak resultant acceleration response of a tall building.

In the pursuit of a better estimate, the use of a partial correlation among the directional maximum component accelerations should be considered. Based on experimental data derived in wind tunnel tests and full-scale measurements taken at the Boundary Layer Wind Tunnel Laboratory, Isyumov et al. (1992) developed empirically the interaction relationship between the maxima of two independent component responses or commonly referred to as the joint action factor $\varphi$. In general, the joint action factor has a value ranging between 0.7 and 1, depending on the ratio of the smaller component to the larger component response. A larger value of the ratio of two component responses implies a smaller value of the joint action factor, or vice versa. A more accurate resultant peak acceleration of a tall building can be computed by multiplying together the joint action factor and the fully correlated value of the peak resultant acceleration response as

$$\hat{a}_{r,\tau} \approx \varphi \sqrt{\hat{a}_{cx,\tau}^2 + \hat{a}_{cy,\tau}^2}$$
$$= \varphi g \sqrt{E\left(a_r^2\right)} = \varphi g \sqrt{\sigma_{\ddot{x}}^2 + \sigma_{\ddot{y}}^2 + \left(R_x^2 + R_y^2\right)\sigma_{\ddot{\theta}}^2} \qquad (5.27)$$

## 5.4 Acceleration response and occupant comfort criteria

Due to the close relationship between the perception of wind-induced motion and the acceleration response of buildings, the checking procedure for motion perception can be conveniently performed by comparing the magnitude of wind-induced acceleration and the acceptability threshold of motion (or the so-called, occupant comfort criteria). Much research effort has been devoted to investigate human perception and tolerance thresholds of wind-induced building motion since 1970s (Hansen et al. 1973; Irwin 1978; Lee 1983; Isyumov et al. 1988; Goto et al. 1990; Melbourne and Palmer 1992; Isyumov and Kilpatrick 1996). Although some recommendations on acceleration limitations have been given in a number of current design codes, there is still no internationally well accepted standard for limiting wind-induced motion in buildings. The research on perception evaluation of wind-induced vibration in buildings is still an active research topic in wind engineering (Tamura et al. 2006; Kwok et al. 2007; Burton et al. 2006, 2007).

There has long been debate about whether human response to wind-induced motion is dependent on the frequency of motion. Chang (1967, 1973) was one of the first to suggest frequency independent occupant comfort criteria based on observations on actual buildings. The first codified recommendations suggesting frequency independent peak acceleration criteria were given in the National Building Code of Canada 1977. Based on the review of a large amount of work (Chen and Robertson 1973, Hansen et al. 1973) regarding many aspects of human response to vibration, field measurements and interviews of occupants, Irwin (1978) showed frequency dependence of uniaxial sinusoidal motion and recommended frequency dependent acceleration criteria for evaluating low-frequency motion within the range of 0.063 Hz to 1.0 Hz, which later led to the development of the ISO-6897 guidelines (ISO 1984). The ISO-6897's motion perception threshold expressed in terms of standard deviation acceleration for a 10-minute duration in 5-year-recurrence wind is given as a function of frequency as

$$\sigma_{\ddot{q}_B}^U = \exp(-3.65 - 0.41 \ln f_j) \qquad (5.28)$$

where $\sigma_{\ddot{q}_B}^U$ denotes the upper limits of the standard deviation modal acceleration response of a building.

In current design practice, it appears that standard deviation acceleration and peak acceleration are two important measures for assessing human perception of building motion. Since the human perception may be more dominated by the overall averaged effect for long duration events of a stationary vibration, using the standard deviation acceleration may be a reasonable choice to check the occupant comfort performance of buildings under more frequently occurring wind events (Boggs 1995; Burton et al. 2007). But for short duration infrequently occurring extreme wind events, such as thunderstorms and typhoons, averaging the acceleration over the worst 10-minutes of the non-stationary motion of buildings may be statistically meaningless, and may also underestimate the effects of the largest individual peaks, whereby occupants are likely in alarm with fear for safety (Burton et al. 2007). It seems that the two kinds of acceleration response (peak acceleration and standard deviation acceleration) may characterize different motion effects on human response. The choice of a particular acceleration criterion for a building may depend on the type of dominating wind events in a geographical location, where the building is to be situated.

Although recent studies have indicated that the evaluation of occupant comfort in buildings should include frequency dependency of motion (Burton et al. 2006, 2007), the incorporation of the frequency dependent modal acceleration criteria into the serviceability design of tall buildings involving complex modes of vibration is problematic. For asymmetric tall buildings with complex 3D mode shapes, the acceleration response should constitute the combination of all significant modal acceleration responses as shown in Eq. (9). While it may be more accurate to include all contributory modes of vibration in the prediction of wind-induced acceleration response of the building, it is not clear, however, whether the acceleration response should be checked against the modal acceleration criteria based on the first mode or a combination of all contributory modes. Furthermore, it is questionable whether the total motion resulting from all contributory modes of vibration may be fully sensed by general occupants. As opposed to the frequency dependent modal acceleration criteria, the frequency independent peak acceleration criteria provide a convenient alternative for checking the acceptability of wind-induced motion in complex tall buildings. Once the resultant acceleration response taking full account of combined swaying and torsional effects at the critical location of a building is determined, its value can then be conveniently checked against a fixed acceleration threshold.

Upon the establishment of wind-induced drift and acceleration criteria, the optimal serviceability design problem can then be mathematically formulated in terms of design variables. In assessing the serviceability performance of modern complex tall buildings, it is necessary to develop a general design optimization framework, which encompasses various kinds of static and dynamic design performance constraints. Various occupant comfort design criteria ranging from modal acceleration criteria corresponding to more frequently occurring wind conditions and peak resultant acceleration criteria associated with less frequently occurring extreme wind events should be included in the design optimization framework as different design requirements under various serviceability performance levels.

## 5.5   Wind-induced performance-based design optimization

Based on a wind and hurricane design framework recommended by Chock et al. (1998), multiple design wind hazard levels as shown in Table 2.1 could be explicitly defined by specific probabilities of exceedance to cover a greater spectrum of possible extreme wind events that threaten to wind-sensitive building structures. When applied to wind-resistant design, performance-based design not only can address issues of occupant comfort and drift serviceability for relative frequent wind environmental conditions, but also treat the rare extreme events for minimizing any possible future damage. In this study, the frequency and occasional levels of wind hazards have been formerly treated by the numerical performance-based optimization technique with the assumption that under these wind events structural responses are kept in the linear-elastic stage. It is noted that the rare extreme wind of 475-year or even 1000-year return period may cause nonlinear-inelastic responses of building structures. Therefore, nonlinear dynamic analysis or an inelastic "pushover" analysis has to be used in the future development to cover the design against the rare extreme wind hazard. The performance-based design optimization problem of tall buildings would then be formulated with various performance design constraints in order to satisfy different performance design objective at specified wind-related performance levels.

Consider a general tall building having $i_s = 1,2, ..., N_s$ steel frame elements, $i_c = 1,2, ..., N_c$ concrete frame elements and $i_w = 1,2, ..., N_w$ concrete shear wall elements. The design problem of minimizing the structural material cost of a tall building subject to the wind-induced drift and acceleration performance design criteria can be posed as:

Minimize

$$W(A_{i_s}, B_{i_c}, D_{i_c}, t_{i_w}) = \sum_{i_s=1}^{N_s} w_{i_s} A_{i_s} + \sum_{i_c=1}^{N_c} w_{i_c} B_{i_c} D_{i_c} + \sum_{i_w=1}^{N_w} w_{i_w} t_{i_w} \qquad (5.29)$$

Subject to

$$d_l \leq d_l^U \qquad (l = 1,2,...,N_l) \qquad (5.30)$$

120

$$\hat{a}_l \le a_l^U \qquad (l = 1,2,...,N_l) \qquad\qquad (5.31)$$

$$\sigma_{\ddot{q}_{jj}} \le \sigma_{\ddot{q}_{jj}}^U \qquad (j = 1,2,...,n) \qquad\qquad (5.32)$$

$$g_j \sigma_{\ddot{q}_{jj}} \le a_{\ddot{q}_{jj}}^U \qquad (j = 1,2,...,n) \qquad\qquad (5.33)$$

$$A_{i_s}^L \le A_{i_s} \le A_{i_s}^U \qquad (i_s = 1,2,...,N_s) \qquad\qquad (5.34)$$

$$B_{i_c}^L \le B_{i_c} \le B_{i_c}^U \qquad (i_c = 1,2,...,N_c) \qquad\qquad (5.35)$$

$$D_{i_c}^L \le D_{i_c} \le D_{i_c}^U \qquad (i_c = 1,2,...,N_c) \qquad\qquad (5.36)$$

$$B_{i_w}^L \le B_{i_w} \le B_{i_w}^U \qquad (i_w = 1,2,...,N_w) \qquad\qquad (5.37)$$

Eq. (5.29) defines the objective function of the structural material cost, in which $w_{i_s}, w_{i_c}, w_{i_w}$ =the respective unit cost of the steel sections, concrete sections and shear walls; $A_{i_s}$ = cross section area of steel section $i_s$; $B_{i_c}, D_{i_c}$ = width and depth of rectangular concrete section $i_c$; and $t_{i_w}$ = thickness of concrete wall section $i_w$. Eq. (5.30) represents the topmost story deflection and interstory drift constraints under 50-year return period wind, in which $l$ denotes the different incident wind angle conditions and $d_l^U$ = the serviceability limit for wind-induced drift responses of tall buildings. In general, the allowable wind-induced drift ratio for tall buildings appears to be within the range of 1/750 to 1/250, with 1/400 being typical. Eq. (5.31) represents the set of $l = 1,2,..., N_l$ peak resultant acceleration performance constraints under $l = 1,2,...,N_l$ different incident wind angle conditions of a typical 10-year return period wind, where $\hat{a}_l$ = the expected maximum peak resultant acceleration; and $a_l^U$ = the corresponding $l$-th predefined peak resultant acceleration limiting value. Eq. (5.32) gives the dynamic serviceability constraints in terms of the standard deviation modal acceleration according to the ISO-6897 criteria for a 10-minute duration in 5-year return period wind. Eq. (5.33) represents the peak modal acceleration constraints based on the AIJ-GEH criteria for a 10-minute duration in 1-year period return wind, where $g_j$ denoting a modal peak factor can be obtained from Eq. (5.23). Eqs. (5.34) to (5.37) define the element sizing bounds in which superscript $L$ denotes lower size bound and superscript $U$ denotes upper size bound of member $i$.

For practical design purpose, wind-induced dynamic loading could be treated as equivalent static loading. The equivalent static load approach takes advantage of the well-established static response analysis and optimization techniques. As described in Chapter 4, the equivalent static wind loads can be obtained and expressed in terms of the alongwind, crosswind and torsional directional loads. Each directional loading consists of the mean, the background and the resonant components. A comprehensive framework for determining the ESWLs on tall buildings could be referred to Chapter 4 as well as the recent literature (e.g., Chen and Kareem 2005a; Chan et al. 2007). A major difference between optimizing for static loads and equivalent static wind loads is the fact that ESWLs are indeed dependent upon the natural frequency of the building. Since the natural frequency of a building is a function of its structural stiffness and mass, making a change in the element sizes of the building may also lead to a change in the value of the ESWLs. In order to make an accurate prediction of the wind-induced drift response under the actions of ESWLs at the time of the structural design phase, it is highlighted in Chapter 4 that the ESWLs be always updated whenever there exists a significant change in the structural properties of the building.

## 5.6  Explicit formulation of acceleration constraints

To facilitate a numerical solution of the design optimization problem, it is necessary that the implicit drift and acceleration performance design constraints through Eq. (5.30) to Eq. (5.33) be formulated explicitly in terms of design variables. Using the principal of virtual work, the elastic drift response of a building under the actions of ESWLs can be explicitly expressed as Eq. (4.31), which is shown in Chapter 4.

Based on the random vibration theory, the standard deviation as well the peak resultant acceleration responses of a wind-excited building can be calculated based on the wind tunnel derived aerodynamic wind load spectra which can be explicitly related to the modal frequency of the structure (Chan and Chui 2006). Substituting Eq. (5.12) into Eq. (5.27) and rearranging terms, the expected maximum peak resultant acceleration could be written as

$$\hat{a}_{R,\tau} = \varphi g \sqrt{\sum_{j=1}^{n} \sum_{s=x,y,\theta} \tilde{\phi}_{js}^2 \sigma_{\ddot{q}_{jj}}^2 + \sum_{j=1}^{n} \sum_{k=1}^{n} \sum_{\substack{s=x,y,\theta \\ j\neq k}} \tilde{\phi}_{js} \tilde{\phi}_{ks} r_{jk} \sigma_{\ddot{q}_{jj}} \sigma_{\ddot{q}_{kk}}} \qquad (5.38)$$

where the transformed swaying mode shapes at the most distant corner of the top floor of the building are given as

$$\tilde{\phi}_{jx} = \phi_{js}(H); \ \tilde{\phi}_{jy} = \phi_{jy}(H); \ \tilde{\phi}_{j\theta} = \sqrt{R_x^2 + R_y^2} \times \phi_{j\theta}(H) \qquad (5.39)$$

where $H$ = the height of the building and $(R_x, R_y)$=planar position of the most distant corner at the top floor of the building.

With the aid of piece-wise regression analysis, the PSD function of modal wind forces for a typical tall building can be inversely related to the modal frequency of the building and is expressed as an algebraic function of the frequency within the typical range of frequency for dynamic serviceability check as follows (Chan and Chiu 2006)

$$S_{Q_{jj}}(f_j) = \beta_j f_j^{-\alpha_j} \qquad (5.40)$$

where $\alpha_j$ and $\beta_j$ are regression constants and normally $\alpha_j > 1$ and $\beta_j > 0$. It is worth to note that some tall buildings with well organized aerodynamic shapes may result in flat modal force spectra with $\alpha_j \approx 0$. In such instances, the modal force spectra and in turn the dynamic responses of these buildings become less sensitive to the modification of building stiffness and some other measures, such as adding mechanical dampers, may become necessary to mitigate wind-induced vibration of buildings. Substituting Eq. (5.40) into Eq. (5.13) gives

$$\sigma_{\ddot{q}_{jj}}^2 \approx \frac{\pi}{4\xi_j m_j^2} \beta_j f_j^{-(\alpha_j - 1)} \qquad (5.41)$$

For wind sensitive tall buildings where the value of the exponent $\alpha_j$ is normally greater than 1, the modal acceleration response can be reduced by increasing modal frequency according to Eq. (5.41). Substituting Eq. (5.41) into Eq. (5.38), the expected maximum peak resultant acceleration can be explicitly expressed in terms of $n$ number of modal frequencies as

123

$$\hat{a}_{R,\tau} \approx g\frac{\sqrt{\pi}}{2}\left[\sum_{j=1}^{n}\sum_{s=x,y,\theta}\tilde{\phi}_{js}^2\frac{\beta_j f_j^{-(\alpha_j-1)}}{\xi_j m_j^2}+\sum_{j=1}^{n}\sum_{\substack{k=1\\j\neq k}}^{n}\sum_{s=x,y,\theta}\tilde{\phi}_{js}\tilde{\phi}_{ks}\frac{r_{jk}}{m_j m_k}\sqrt{\frac{\beta_j\beta_k f_j^{-(\alpha_j-1)}f_k^{-(\alpha_k-1)}}{\xi_j\xi_k}}\right]^{1/2}$$

(5.42)

Using the Rayleigh Quotient method, the modal frequency $f_j$ or natural period $T_j$ of a building system can be related to the total internal strain energy $U_j$ of the structural system due to the $j$-th modal inertia force applied statically to the system as follows (Huang and Chan, 2007):

$$U_j = c_j \big/ f_j^2 = c_j T_j^2 \qquad (5.43)$$

where $c_j$ denotes a proportionality constant connecting the internal strain energy $U_j$ to the square of the modal frequency $f_j$ or the square of the nature period $T_j$ of the system. By definition, the internal strain energy $U_j$ could be calculated based on the inertial-induced internal force of each structural member.

Substituting Eq. (5.43) into Eq. (5.42), the occupant comfort peak resultant acceleration performance constraints given in Eq. (5.31) can be expressed as a function of the total internal strain energy of the system $U_j$ and then successively in terms of the element sizing design variables as follows:

$$\hat{a}_l = g\frac{\sqrt{\pi}}{2}\left[\sum_{j=1}^{n}\sum_{s=x,y,\theta}\tilde{\phi}_{js}^2 A_j\left(U_j\right)^{\frac{\alpha_j-1}{2}}+\sum_{j=1}^{n}\sum_{k=1}^{n}\sum_{\substack{s=x,y,\theta\\j\neq k}}\tilde{\phi}_{js}\tilde{\phi}_{ks}B_{jk}\left(U_j\right)^{\frac{\alpha_j-1}{4}}\left(U_k\right)^{\frac{\alpha_k-1}{4}}\right]^{1/2}\leq a_l^U \qquad (5.44)$$

where the constants $A_j$ and $B_{jk}$ can be given as

$$A_j = \frac{\beta_j}{\xi_j m_j^2}\left(\frac{1}{c_j}\right)^{\frac{\alpha_j-1}{2}}\;;\quad B_{jk} = \frac{r_{jk}}{m_j m_k}\sqrt{\frac{\beta_j\beta_k}{\xi_j\xi_k}}\left(\frac{1}{c_j}\right)^{\frac{\alpha_j-1}{4}}\left(\frac{1}{c_k}\right)^{\frac{\alpha_k-1}{4}} \qquad (5.45)$$

Similarly, both the standard deviation modal acceleration and peak modal acceleration values given in the occupant comfort performance constraints of Eqs. (5.32) and (5.33) can also be expressed by the modal internal strain energy of the system $U_j$ and in turn by the element sizing variables as follows:

124

$$\sigma_{\ddot{q}_{jj}} \approx \left[ \frac{\pi}{4\xi_j m_j^2} \beta_j \left( \frac{c_j}{U_j} \right)^{\frac{\alpha_j-1}{2}} \right]^{1/2} = \frac{\sqrt{\pi}}{2} \left[ A_j (U_j)^{\frac{\alpha_j-1}{2}} \right]^{1/2} \qquad (5.46)$$

$$g_j \sigma_{\ddot{q}_{jj}} \approx \left[ \sqrt{\ln \tau^2 \left( \frac{c_j}{U_j} \right)} + 0.5772 / \sqrt{\ln \tau^2 \left( \frac{c_j}{U_j} \right)} \right] \frac{\sqrt{\pi}}{2} \left[ A_j (U_j)^{\frac{\alpha_j-1}{2}} \right]^{1/2} \qquad (5.47)$$

The frequency dependent acceleration criteria involved in Eq. (5.32) and (5.33) could also be related to the modal internal strain energy as

$$\sigma_{\ddot{q}_{jj}}^U = \exp(-3.65 - 0.41 \ln f_j) = \exp\left[ -3.65 - 0.205 \ln \left( \frac{c_j}{U_j} \right) \right] \qquad (5.48)$$

The AIJ-GEH's peak modal acceleration criteria for office buildings involved in Eq. (5.33) can be approximated by the following expression (Kwok et al. 2007) as

$$a_{\ddot{q}_{jj}}^U = 3.5*0.736*\exp(-3.65 - 0.41 \ln f_j) = 3.5*0.736*\exp\left[ -3.65 - 0.205 \ln \left( \frac{c_j}{U_j} \right) \right] (5.49)$$

where the value of 3.5 is a typical value of the peak factor; 0.736 is a multiplication factor, which converts the limiting acceleration from a recurrence interval of five years to one year.

For a tall building of mixed steel and concrete construction, the total internal strain energy of the building structure due to the $j$-th modal inertia force can be obtained by summing up the internal work done of each structural member as:

$$\left. \begin{aligned} U_j(A_i, B_i, D_i) = \sum_{i_s=1}^{N_s} \left( \frac{e_{i_s j}}{A_{i_s}} + e'_{i_s j} \right) + \sum_{i_c=1}^{N_c} \left( \frac{e_{0i_c j}}{B_{i_c} D_{i_c}} + \frac{e_{1i_c j}}{B_{i_c} D_{i_c}^3} + \frac{e_{2i_c j}}{B_{i_c}^3 D_{i_c}} \right) \\ + \sum_{i_w=1}^{N_w} \left( \frac{e_{0i_w j}}{t_{i_w}} + \frac{e_{1i_w j}}{t_{i_w}^3} \right) \qquad (j=1,2,...n) \end{aligned} \right\} \qquad (5.50)$$

where $e_{i_s j}, e'_{i_s j}$ = the internal strain energy coefficients and its correction factor of steel member $i_s$;

$e_{0i_c j}, e_{1i_c j}, e_{2i_c j}$ = the internal strain energy coefficients of concrete member $i_c$; and $e_{0i_w j}, e_{1i_w j}$ the internal strain energy coefficients of concrete wall section $i_w$. Note that the total strain energy expression (Eq. (5.50)) relating to the modal frequency of the building system looks very similar to the virtual strain energy expression (Eq. (4.31)) relating to the static wind drift response of the system. The only difference is that the strain energy expression given in Eq. (5.50) is derived from the static analysis of the structure under the application of the modal inertia forces, whereas the energy expression Eq. (4.31) is derived from the static analysis of the structure due to the application of the actual ESWLs and virtual load cases.

Upon establishing the explicit formulation of the drift and acceleration performance design constraints, the next task is to apply a suitable numerical technique for solving the optimal performance-based design problem. A rigorously derived Optimality Criteria (OC) method, as described in Chapter 4, is herein employed. The overall procedure of wind-induced performance-based design optimization of tall buildings is presented by the flow chart as shown in Figure 5.1.

## 5.7    Illustrative example

### 5.7.1    The 60-story benchmark building with 4-story height outriggers

A 60-story building having a height of 240m and a rectangular floor plan dimension of 24 m by 72 m is used to illustrate the effectiveness and practical application of the optimal serviceability design technique. As shown in Figures 5.2, the building has adopted basically an outrigger braced system, which consists of a perimeter steel frame connected to an eccentric reinforced concrete core by two levels of 4-story steel outriggers and belt trusses. Two double-bay X-braced frames are also placed at the two respective end faces of the building to enhance the torsion rigidity of the building. Since the concrete lift core is eccentrically located, significant ample lateral-torsional motion induced by wind loading is anticipated.

The exterior steel frames of the building were to be designed using AISC standard steel sections as follows: W30 shapes for beams; W14 shapes for columns, braces, belt and outrigger trusses. The eccentric core wall of the building is made of grade C40 concrete. The initial structural element sizes of the building established based on a preliminary strength check are given in Table 5.1. Once the finite element model of the building was created, an eigenvalue analysis was then executed to determine the dynamic properties of the building (i.e., the natural frequencies and mode shapes). The three-dimensional mode shapes of the building are given in Figure 5.3. The first three fundamental vibration modes of the building have corresponding natural frequencies of 0.185 Hz (swaying primarily in the short direction of the building), 0.327 Hz (swaying primarily in the long direction) and 0.410Hz (mainly torsional vibration). The modal damping ratio of 1% and 1.5% for the first three fundamental vibration modes of the building were assumed for calculating the acceleration and drift responses, respectively.

As described in Chapter 3, the wind tunnel test was carried out at the CLP Power Wind/Wave Tunnel Facility (WWTF) of the Hong Kong University of Science and Technology (Tse et al. 2007). Two specific wind conditions corresponding to two perpendicular incident wind angles were simultaneously considered in the design optimization synthesis. One was the 0-degree wind perpendicular to the wide face acting in the short direction (i.e. along the Y-axis) of the building; another one was the 90-degree wind perpendicular to the narrow face acting in the long direction (i.e. along the X-axis). The PSD functions of modal forces corresponding to a 10-year return period wind for the building are presented in Figure 5.4. For the incident wind angle of 0 degree with wind normal to the wide face of the building, the wind-induced modal force spectra with the typical range of frequency for dynamic serviceability check (i.e. $0.1\text{Hz} \leq f \leq 1\text{Hz}$ ) can be expressed as a function of the modal frequency $f$ as follows:

Mode 1: $\qquad S_{Q_{11}} = 4.660 \times 10^8 f^{-4.069}$ (N²/Hz) $\qquad$ (5.51)

Mode 2: $\qquad S_{Q_{22}} = 2.467 \times 10^8 f^{-3.808}$ (N²/Hz) $\qquad$ (5.52)

Mode 3: $\qquad S_{Q_{33}} = 3.437 \times 10^8 f^{-3.835}$ (N²/Hz) $\qquad$ (5.53)

For the incident wind angle of 90 degree with wind normal to the narrow face of the building, the wind-induced modal force spectra can be written as:

Mode 1: $\qquad S_{Q_{11}} = 7.617 \times 10^8 f^{-3.879}$ (N²/Hz) $\qquad$ (5.54)

Mode 2: $\qquad S_{Q_{22}} = 2.490 \times 10^8 f^{-3.096}$ (N²/Hz) $\qquad$ (5.55)

Mode 3: $\qquad S_{Q_{33}} = 7.874 \times 10^8 f^{-4.717}$ (N²/Hz) $\qquad$ (5.56)

For other recurrence intervals of wind, the modal force spectra corresponding to a given wind speed $V_R$ in the recurrence interval of $R$ years can be obtained by multiplying Eqs. (5.51) to (5.56) associated with the 10-year-recurrence wind speed $V_{10}$ by a factor $\left( V_R / V_{10} \right)^4$.

### 5.7.2   Site-specific design wind speed estimation

The most valuable source of information in the estimates of future extreme wind speeds at the building project site is the historical wind speed data that has been recorded at the building site or at a nearby observatory site. The wind speed data used in this paper has been published on line by Hong Kong Observatory. The basic reference wind speeds adopted by the Hong Kong Wind Code

2004 for the establishment of the design wind profile are derived from extreme wind analysis of the measured wind data at Waglan Island, Hong Kong. All available typhoon data measured at Waglan Island since 1953 to 2006 form the basis for wind speed analysis in this paper. 110 data points of observed typhoon wind speeds with a magnitude greater than 16 m/s has been used in the statistical analysis. The anemometer height at Waglan Island changed from 75m before 1993 to 83m after 1993. Taking into account the steep rocky profile of the island which presents a blockage to the wind and speed up effect over the island, the measured data are considered corresponding to an anemometer height of 90m.

The extreme wind speed analysis was done using the probability paper method as presented in section 5.2. Figure 5.5 presents the Gumbel plot of typhoon hourly-mean wind speeds at a reference height of 90m. The design wind speed results for 10-year, 50-year and 475-year return periods give hourly-mean wind speeds of 34.7m/s, 43.6m/s and 55.7m/s, respectively. According to the Hong Kong Wind Code 2004, design hourly-mean wind speed at reference height of 90m for 50-year return period is 46.9m/s, which is greater than 43.6m/s predicted in this paper. The main reason of inconsistent wind speed estimates is that the wind data used in the wind speed predictions for the Hong Kong Wind Code 2004 was only covered from 1953-1999. When applying the same wind analysis procedure to the typhoon data year range of 1953-1999, the design wind speed result for 50-year return periods then gives hourly-mean wind speeds of 47.1m/s, which is very close to 46.9 m/s adopted by the Hong Kong code. It is noticed that the discrepancy between 47.1m/s and 46.9m/s is within a slight difference of 0.5% due to sampling errors. The decreasing of design wind speeds resulted from the inclusion of the recent typhoon wind speed data indicates that the magnitudes of typhoon events occurred during the recent 7 years tends to be relatively small.

The uncertainties in the estimation of design wind speeds for 50-year return period were quantified in terms of coefficients of variation based on the distribution parameters of Type I extreme value. The coefficient of variation could be expressed as a function of the sample size $N$. As shown in Figure 5.6, the value of COV is monotonically reduced as the total number of sample data increasing. The sampling error due to using $N=110$ was quantified as 4.4%, which indicates that the design wind speed for 50-year return period would be within the range of (43.6±4.4%) m/s. According to the different acceleration design constraints given in Eq. (5.31) to (5.33), 10-year, 5-year and 1-year return period of winds in a typhoon-prone urban environment like Hong Kong were considered for evaluating acceleration performances of the building, respectively. Following the Hong Kong Wind Code 2004, the design hourly-mean wind speeds at reference height of 90m corresponding to the three different return periods of wind are given in Table 5.2.

### 5.7.3 Results and discussion

The component and resultant acceleration responses of the building are given in Table 5.3. Due to vortex shedding effects, significant crosswind vibrations in the Y-direction (short direction) induced by the 90-degree wind is found. For instance, the Y-direction crosswind acceleration component induced by the 90-degree wind on the building is 34.9 milli-g, which is larger than the corresponding Y-direction alongwind acceleration components of 28.7 milli-g under the 0-degree wind. By combining the swaying and torsional acceleration components, the resultant accelerations at the most distant corner of the top floor level of the building are 33.5 milli-g under the 0-degree wind and are 44.3 milli-g under the 90-degree wind as shown in Table 5.3. Significant violations of 34% and 77% in the peak resultant acceleration criterion of 25 milli-g according to the Hong Kong Codes of Practice (HKCOP 2004, HKCOP 2005) are found for the building under the 0-degree wind and the 90-degree wind, respectively.

For comparison, the habitability or occupant comfort of the initial building under more frequent 5-year and 1-year return period wind was also assessed based on the frequency dependant ISO-6897 and AIJ-GEH criteria. The modal acceleration responses of the building and the frequency dependant threshold values stipulated in the ISO-6897 and AIJ-GEH are also given in Table 5.3. Before optimization, the building is found to violate the ISO-6897 occupant comfort criteria for the first modal response, but to meet the ISO-6897 criteria for the second and third modes. In terms of standard deviation modal acceleration, the first modal responses of the building violated the ISO-6897 criteria by 8% and 19% under the 0-degree wind and the 90-degree wind, respectively. In terms of peak modal accelerations, the initial building structure is found to meet the AIJ-GEH criteria for all three modes under both the 0-degree and 90-degree wind conditions.

For this 60-story building with 3D coupled modes, the frequency independent peak resultant acceleration criteria appear to be most critical, while the ISO-6897 frequency dependent standard deviation modal acceleration criteria seem to be stricter than the AIJ-GEH peak modal acceleration criteria. Since the peak resultant acceleration response captures not only the mechanically coupled swaying and torsional effects but also the modal combination effects including torsional amplification at the corner positions, designing the building to satisfy the resultant acceleration criteria is thus more challenging than that the modal acceleration criteria of the ISO-6897 or AIJ-GEH. In order to study the influence of the occupant comfort criteria on the optimal stiffness design of the building, four optimization cases with different consideration of drift and acceleration constraints are considered:

129

Case 1: consideration of drift constraints only.

Case 2: drift constraints and the ISO-6897 standard deviation modal acceleration constraints.

Case 3: drift constraints and the AIJ-GEH peak modal acceleration constraints.

Case 4: drift constraints and the HKCOP peak resultant acceleration constraints.

Figure 5.7 presents the material cost design histories of the building for all four cases. The normalized cost with respect to the initial structural material cost of the building is given for each design cycle, which includes the process of one formal response analysis and one resizing optimization. Although the structural cost of the building was found somewhat fluctuating for the first few design cycles, steady convergence to the final optimum solution was achieved for all 4 cases. Cases 1, 2 and 3 have converged to almost the same value of the normalized cost (i.e., 0.962 for case 1, 0.971 for case 2 and 0.958 for case 3). Such a result and a subsequent check on the final element sizes of the optimized structures indicate that the standard deviation modal acceleration constraints in case 2 and the peak modal acceleration constraints in case 3 have relatively little influence on the optimized building structure. It is evident that the optimized designs of the first three cases are mainly governed by the static drift constraints rather than by the dynamic modal acceleration constraints. When both static drift and peak resultant acceleration constraints are considered in case 4, an increase of about 21.5% in the material cost is needed to fulfill both the drift and peak resultant acceleration constraints, indicating that the peak resultant acceleration has greatly influenced on the final design of the structure.

Table 5.4 presents the acceleration responses of the optimized building and the evaluation of habitability after optimization. The modal acceleration responses of the optimized structure are found to be within the ISO-6897 standard deviation modal acceleration limit for case 2, and within the AIJ-GEH peak modal acceleration limit for case 3. In case 4, the initial peak resultant response of 44.3 milli-g is greatly reduced by 56% to slightly below the limiting value of 25 milli-g. The component and resultant acceleration responses at the top corner of the building after optimization are also given in Table 5.4. It is found that all three acceleration components have been reduced by the stiffness design optimization technique. The most significant reduction of acceleration is the acceleration component caused by the torsional effect. For example, the torsional acceleration components at the top corner of the building are reduced by 60.1% (from 15.3 to 6.1 milli-g) and 67.1% (from 27.1 to 8.9 milli-g) under the 0-degree wind and the 90-degree wind, respectively. It is believed that the great reduction in the torsional acceleration is due to the significant improvement in the torsional rigidity of the building by the computer based optimization technique.

Table 5.5 presents modal frequencies for the optimized building with different optimization cases. For case 1, 2 and 3, the optimized modal frequency values for three modes are quite close to each other. The similar optimized modal frequency results is due to that the drift constraints are the dominating design constraints rather than the standard deviation or peak modal acceleration constraints during the optimization processes of case 2 and 3. In order to meet stricter peak resultant acceleration criteria, 4% to 9% higher values for three modal frequencies are needed for the optimum design solutions in case 4 compared with all the other cases as shown in the table. After optimization, in case 4 the torsional stiffness is much more enhanced than the lateral stiffness such that the third modal frequency is more increased by 9% than 4% increase for the first modal frequency.

Figures 5.8 and 5.9 present the initial and final lateral deflections profiles and interstory drift ratio profiles at the most critical corner column of the building for case 4. It is clearly shown that the initial design established solely based on strength check are found to violate considerably in both the top deflection limit (i.e. H/400, where H = building height) and interstory drift ratio limit (i.e. 1/400). With a wider frontal face in the 0-degree wind, much larger lateral deflection and interstory drift responses in the short direction (i.e. along the Y-axis) of the building are found as shown in Figures 5.8 and 5.9. After the optimization, no violation in both the top deflection and the interstory drift responses are found for the optimized building under the 0-degree wind and the 90-degree wind. It is noted that the two kinks on each interstory drift ratio profile as shown in Figure 7 indicate the two levels of the steel outriggers and belt trusses.

## 5.8 Summary

The stiffness optimization method for drift design presented in Chapter 4 is extended to the optimal performance-based dynamic serviceability design. The performance-based design optimization technique is proposed for automatic optimal design satisfying simultaneously the requirements imposed by the different performance levels associated with the performance design objectives (design criteria). Performance-based design wind speed with quantitative uncertainty at various wind hazard severity level is obtained by performing statistical analysis on extreme wind speed data.

Specifically, an integrated wind-induced dynamic analysis and automatic computer-based design optimization technique is developed for minimizing the structural cost of tall buildings subject to dynamic performance design criteria. Using experimentally derived aerodynamic wind load spectra, the coupled dynamic response of a tall building under spatiotemporally varying wind excitation is formulated based on random vibration theory. The drift constraints and the occupant comfort acceleration performance design constraints have been explicitly expressed in terms of the design variables. A rigorously derived Optimality Criteria (OC) method has been developed to solve for the optimal stiffness design solution of a tall building satisfying the wind-induced drift and acceleration design constraints. Results of a full-scale 60-story hybrid building with 3D mode shapes have shown that the stiffness design optimization technique provides a powerful design tool for the drift and occupant comfort serviceability design of wind-excited tall buildings. Not only is the computer-based optimization technique capable of achieving the most economical distribution of element stiffness of practical tall building structures while satisfying multiple static drift and dynamic acceleration performance design criteria, but also the numerical optimal design method is computationally efficient since the final optimal design can generally be obtained in only a few number of reanalysis and redesign cycles. Both frequency independent peak resultant acceleration criteria and frequency dependent standard deviation or peak modal acceleration criteria have been considered in the optimization process. It was found that various occupant comfort acceleration criteria have different implication on the evaluation of habitability and the final optimal design solution does depend on the particular choice of occupant comfort criteria.

**Table 5.1    Initial member sizes for the 60-story benchmark building**

| Concrete core | | Exterior steel framework | | | X Brace and truss | |
|---|---|---|---|---|---|---|
| Concrete wall | | AISC standard steel section | | | X Brace at two end faces | Two levels of outriggers and belt trusses |
| Story no. | Thickness (mm) | Story no. | Column | Beam | | |
| 51-60 | 300 | 57-60 | W14×34 | W30×99 | W14×90 | |
| 41-50 | 600 | 53-56 | W14×34 | W30×99 | W14×90 | |
| 31-40 | 600 | 49-52 | W14×34 | W30×99 | W14×159 | |
| 21+30 | 800 | 45-48 | W14×159 | W30×357 | W14×159 | |
| 11-20 | 800 | 41-44 | W14×159 | W30×357 | W14×159 | W14×730 |
| 1-10 | 1000 | 37-40 | W14×257 | W30×357 | W14×342 | |
| | | 33-36 | W14×257 | W30×357 | W14×342 | |
| | | 29-32 | W14×370 | W30×357 | W14×398 | |
| | | 25-28 | W14×370 | W30×357 | W14×398 | |
| | | 21-24 | W14×370 | W30×357 | W14×398 | |
| | | 17-20 | W14×500 | W30×357 | W14×455 | |
| | | 13-16 | W14×500 | W30×357 | W14×455 | |
| | | 9-12 | W14×730 | W30×357 | W14×730 | W14×730 |
| | | 5-8 | W14×730 | W30×357 | W14×730 | |
| | | 1-4 | W14×730 | W30×357 | W14×730 | |

**Table 5.2   Design wind speed at the height of 90 m in Hong Kong area**

| Design wind speed (m/s) | Average return period | Serviceability check |
|---|---|---|
| 22.5 | 1 year | Occupant perception (AIJ-GEH) |
| 30.6 | 5 years | Occupant comfort (ISO-6897) |
| 34.7 | 10 years | Fear for safety (Hong Kong code) |
| 46.9 | 50 years | Drift (Hong Kong code) |

**Table 5.3   Modal acceleration and peak resultant acceleration of the initial building**

| Criteria | | Mode 1 (milli-g) | Mode 2 (milli-g) | Mode 3 (milli-g) | |
|---|---|---|---|---|---|
| ISO-6897 (1984) 5-year recurrence wind | Standard deviation criteria | 5.2 | 4.1 | 3.8 | |
| | 0-degree wind | 5.6 | 1.1 | 1.1 | |
| | 90-degree wind | 6.2 | 0.8 | 2.4 | |
| AIJ-GHE (2004) 1-year recurrence wind | Peak acceleration criteria | 13.4 | 10.6 | 9.7 | |
| | 0-degree wind | 6.2 | 1.3 | 1.4 | |
| | 90-degree wind | 7.0 | 1.1 | 2.5 | |
| HKCOP (2004) 10-year recurrence wind | Peak resultant acceleration criteria | $a_x$ (milli-g) | $a_y$ (milli-g) | $a_\theta$ ×37.95m[*] (milli-g) | Peak resultant (milli-g) |
| | 0-degree wind | 7.9 | 28.7 | 15.3 | 33.5 |
| | 90-degree wind | 4.9 | 34.9 | 27.1 | 44.3 |

* The distance 37.95 m is measured from the mass center to the corner position of the building plan. Joint action factor is taken as 1.

Table 5.4 Modal acceleration and peak resultant acceleration of the optimized building

| Criteria | | Mode 1 (milli-g) | Mode 2 (milli-g) | Mode 3 (milli-g) |
|---|---|---|---|---|
| ISO-6897 (1984) 5-year recurrence wind | Standard deviation criteria | 4.8 | 3.6 | 3.3 |
| | 0-degree wind | 4.1 | 0.7 | 0.7 |
| | 90-degree wind | 4.6 | 0.5 | 1.3 |
| AIJ-GHE (2004) 1-year recurrence wind | Peak acceleration criteria | 12.3 | 9.2 | 8.5 |
| | 0-degree wind | 4.6 | 0.8 | 0.9 |
| | 90-degree wind | 5.3 | 0.7 | 1.4 |
| HKCOP (2004) 10-year recurrence wind | Peak resultant acceleration criteria | $a_x$ (milli-g) | $a_y$ (milli-g) | $a_\theta$ ×37.95m (milli-g) | Peak resultant (milli-g) |
| | 0-degree wind | 3.9 | 21.6 | 6.1 | 22.8 |
| | 90-degree wind | 2.7 | 23.2 | 8.9 | 24.9 |

Table 5.5 Modal frequencies for the optimized buildings with different cases

| Optimization cases | constraints | Mode 1 (Hz) | Mode 2 (Hz) | Mode 3 (Hz) |
|---|---|---|---|---|
| Case 1 | drift | 0.2270 | 0.4635 | 0.5609 |
| Case 2 | drift and standard deviation modal acceleration | 0.2274 | 0.4631 | 0.5613 |
| Case 3 | drift and peak modal acceleration | 0.2268 | 0.4632 | 0.5611 |
| Case 4 | Drift and peak resultant accelerations | 0.2359 | 0.5046 | 0.6110 |

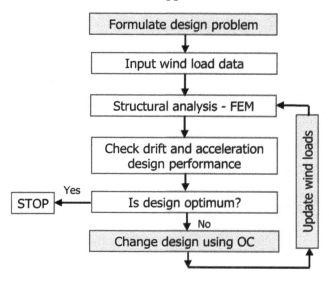

**Figure 5.1** The flow chart of performance-based design optimization of wind sensitive tall buildings

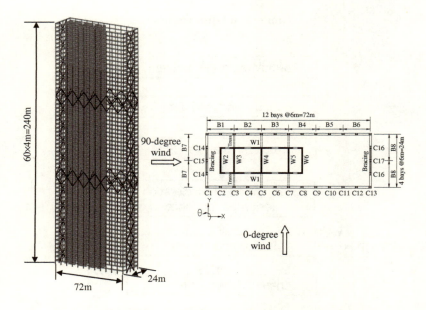

**Figure 5.2    A 60-story benchmark building with 4-story height outriggers**

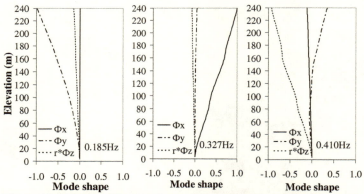

**Figure 5.3    Coupled 3D mode shapes of the 60-story benchmark building**

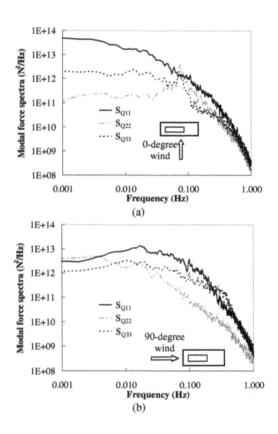

**Figure 5.4    PSDs of modal forces of the 60-story benchmark building**

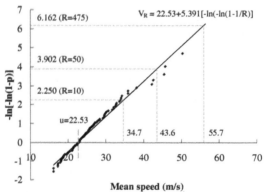

**Figure 5.5    Design hourly-mean wind speed at the height of 90 m in Hong Kong area (1953-2006)**

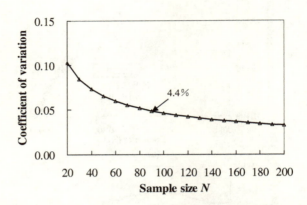

**Figure 5.6    The uncertainty in estimation of design wind speeds for 50-year return period**

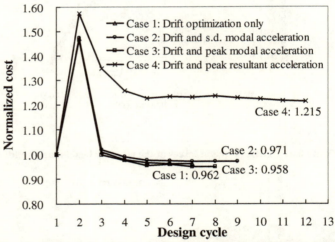

**Figure 5.7    Design histories of the normalized structure cost of the 60-story building**

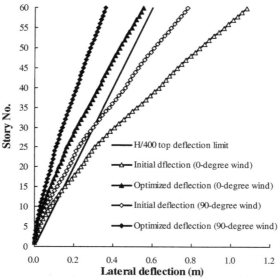

**Figure 5.8    Lateral deflection profiles of the 60-story building for case 4**

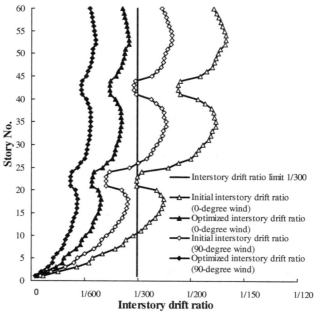

**Figure 5.9    Interstory drift ratio profiles of the 60-story building for case 4**

# CHAPTER 6    Peak Response Statistics and Time-variant Reliability

## 6.1    Introduction

The OC design optimization methodology developed in Chapters 4 and 5 is apparently very promising to solve the wind-induced performance-based design problems of tall buildings. The serviceability design optimization problem addressed in the earlier chapters, however, is deterministic without explicit consideration of uncertainty and reliability issues during the optimization process. Starting from Chapter 6, various uncertainties inherently existed in all aspects related to wind effects on structures are identified, modelled and formerly treated using statistical and reliability analysis method in the further development of reliability performance-based design optimization framework. Moreover, wind-induced dynamic response analysis is to be carried out using the time domain approach instead of the frequency domain approach, as used in Chapters 3 to 5.

In structural design application, it is necessary to estimate the expected maximum peak response, which can be obtained as the standard deviation or RMS multiplied by a peak factor, as demonstrated in Chapter 5 for predicting peak resultant acceleration responses. Davenport (1964) has shown that, if the underlying parent distribution of a process is Gaussian, then the largest values of the process will asymptotically follow a Gumbel distribution. In general, the formulation of Davenport's peak factor provides satisfactory estimate of the expected maximum response for a wideband response process, whereas, for a narrowband response process it yields conservative estimate (Kareem 1987; Gurley et al. 1997). Vanmarcke (1972, 1975) improved the estimation for the first-passage probability by taking into account the dependence between barrier crossings of the narrowband random response process. From the Vanmarcke's results, a probabilistic peak factor could be developed and explicitly expressed as a function of excitation duration, the probability of exceedance and the shape parameter (bandwidth) of the power spectral density (PSD) of the underline random response process. The probabilistic peak factor can be directly related to an explicit measure of time-variant reliability. The results of the probabilistic factor demonstrate that the use of the Davenport's formula in estimating the expected largest response can lead to overly conservative results for a narrowband response process.

Not only does the time history analysis method provide an alternative method to estimate the RMS value of various responses compared to the spectral analysis method, but it also offers much more information on peak response distribution. Based on the data sample of random response process obtained from the numerical time history analysis, response value distribution and the expected values of largest peak response could be statistically estimated (Cheng et al. 2003). From a viewpoint of extremal statistics (Gumbel, 1958), the overall largest peak response could be derived asymptotically from the distribution knowledge of the local peak response. Hence, the asymptotic theory of statistical extremes is applied to determine a peak factor corresponding to a specific peak distribution, which is estimated from the available data samples of a random stationary response process.

Therefore, the objectives of this part of study are (1) to investigate level-crossing rate (LCR) and peak distribution of a random response process; (2) to more appropriately determine peak factors for any given stationary response process and for a combined process based on different PDF models of peak distribution; and (3) to devise new methods to accurately predict the expected maximum response based on the peak factor approach. Finally, the explicit relationship between a peak factor and the time-variant reliability is established.

Specifically, the mathematical analysis of mean LCR for a stationary random process is firstly reviewed in the context of a first-passage problem. From the classical results of the first-passage probability, the analytical expression of the probabilistic peak factor, which is directly related to an explicit measure of reliability (or the probability of random peak responses without exceeding certain threshold value), is obtained. Then, the asymptotic theory of statistical extremes is applied to determine the specific peak factor corresponding to various initial peak distributions. The expected Weibull peak factor is derived from the Weibull peak distribution, which is a generalization of the Rayleigh distribution (Newland 1984). The Gamma peak factor is obtained to facilitate the prediction of the expected maximum response for a resultant response process which follows generally a non-Gaussian distribution. The relationship of peak factors to reliability has also been expressed explicitly given the local maxima distribution of the underlying response process.

Utilizing the wind tunnel data derived from synchronous pressure measurements, time history analysis is carried out for the 60-story benchmark building and the 45-story CAARC building. The usage of the probabilistic peak factor, the Weibull peak factor and the Gamma peak factor are demonstrated. The different methods for predicting the largest peak acceleration responses are compared. The first-order reliability analysis corresponding to different methods is also worked out.

## 6.2   First-passage probability: Poisson models

### 6.2.1 Mean level-crossing rate of stationary random process

#### 6.2.1.1 Stationary Gaussian random process

In the context of random vibration, time-variant reliability has been traditionally posed as the first-passage problem (Vanmarcke 1975). A well known approximation of the first-passage probability of barrier-crossing is obtained by assuming that threshold level crossings occur independently according to a Poisson process. The occurrence of events (crossing the double barrier of level $b$) in two non-overlapping time intervals is stochastically independent, the probability of occurrence of $n$ number of events in the time interval $[0,t]$ follows Poisson model as

$$P_n(t) = \frac{(v_b t)^n}{n!} \exp(-v_b t) \qquad (6.1)$$

where $v_b$ indicates mean out-crossing rate of the level $b$, which is equal to the expected number of exceeding occurrences that occur during the unit time interval. Let $n=0$ in Eq. (6.1), one obtains

$$R(t) = P_0(t) = \exp(-v_b t) \qquad (6.2)$$

where $R(t)$ could be called as a reliability function, which means the probability of no out-crossing during the given time interval. If the random response of interest is stationary, the desired maximum reliability for a given time duration can be attained by minimizing the mean out-crossing rate. The so-called criterion of the minimum failure rate is one of design criteria for mechanical machines and structures operating in a random and harsh environment.

The results of mean level-crossing rate for a Gaussian random response process $Y(t)$ were firstly obtained by Rice (1945). As an alternative approach, Middleton (1960) obtained the same results by introducing a counting function for crossings at the level of $b$. Middleton's approach is mathematically more elegant and provides a unified basis for determining the other statistical properties of a random process, for example, the peak distribution and the first-passage time distribution. The desired counting function may be derived from a zero-one process $Z(t)$

$$Z(t) = u[Y(t) - b] \qquad (6.3)$$

where $u(t)$ is a unit step function, for example, $u(t)=1$ when $t>0$, or $u(t)=0$ when $t<0$. Differentiating $Z(t)$ gives the number of crossings per second as the desired counting function

$$\dot{Z}(t) = \dot{Y}(t)\delta[Y(t) - b] \qquad (6.4)$$

where $\delta(t)$ is a Dirac delta function. The total number of crossings in an interval $(t_2-t_1)$ including both upward and downward passages, is therefore defined as a new random variable $N_b$

$$N_b(t_1, t_2) = \int_{t_1}^{t_2} \left| \dot{Y}(t) \right| \delta[Y(t) - b] dt \qquad (6.5)$$

The mathematical expectation of $N_b$ could be calculated as

$$
\begin{aligned}
E\left[N_b(t_1, t_2)\right] &= \int_{t_1}^{t_2} E\left\{\left|\dot{Y}(t)\right| \delta[Y(t) - b]\right\} dt \\
&= \int_{t_1}^{t_2} \int_{-\infty}^{\infty} \int_{-\infty}^{\infty} \left|\dot{y}\right| \delta(y - b) f_{Y, \dot{Y}}(y, \dot{y}) dy d\dot{y} dt \qquad (6.6) \\
&= \int_{t_1}^{t_2} \int_{-\infty}^{\infty} \left|\dot{y}\right| f_{Y, \dot{Y}}(b, \dot{y}) d\dot{y} dt
\end{aligned}
$$

where $f_{Y, \dot{Y}}(y, \dot{y})$ is the joint probability density function of $Y(t)$ and its derivative $\dot{Y}(t)$. Since the process is stationary, the average number of crossings per second over the interval $(t_2-t_1)$ is equal to the mean level-crossing rate $\nu_b$, which is mathematically defined as the expectation of the number of crossings per second (at an instant $t$). From Eq. (6.6), the well known Rice's formula (1945) is obtained as

$$\nu_b = \int_{-\infty}^{\infty} \left|\dot{y}\right| f_{Y, \dot{Y}}(b, \dot{y}) d\dot{y} \qquad (6.7)$$

Rice's formula could be equally applied for any random response of a linear or non-linear system subjected to Gaussian or non-Gaussian excitation, once the joint probability density function of $f_{Y, \dot{Y}}(y, \dot{y})$ is available. Generally speaking, depending on the properties of system and excitation, the analytical solution of $f_{Y, \dot{Y}}(y, \dot{y})$ is difficult to obtain. Only in a few cases, the exact stationary solutions of response distributions have been obtained, e.g., the output of a linear system driven by Gaussian excitation and the output of a non-linear Duffing oscillator subjected to Gaussian white-noise excitation (Crandall 1963; Nigam 1983). A linear system subjected to Gaussian excitation has been frequently employed to model wind-induced responses of tall buildings by many researchers and engineers. Therefore, it is reasonable to assume that the wind-induced response of $Y(t)$ is Gaussian with zero-mean, then so its derivative $\dot{Y}(t)$ (and so are higher derivatives such as the acceleration process $\ddot{Y}(t)$). Since the two stationary processes of $Y(t)$ and $\dot{Y}(t)$ are independent and uncorrelated, the joint probability density function is readily determined by the joint Gaussian distribution as

143

$$f_{Y,\dot{Y}}(b,\dot{y}) = \frac{1}{2\pi\sigma_Y\sigma_{\dot{Y}}}\exp\left\{-\left(\frac{b^2}{2\sigma_Y^2}+\frac{\dot{y}^2}{2\sigma_{\dot{Y}}^2}\right)\right\} \tag{6.8}$$

where $\sigma_Y$ and $\sigma_{\dot{Y}}$ indicate the RMS value of the response process and its first derivative process, respectively.

Substituting Eq. (6.8) into Eq. (6.7), the mean out-crossing rate of level $b$ for Gaussian process is obtained as

$$\nu_b = \frac{1}{\pi}\frac{\sigma_{\dot{Y}}}{\sigma_Y}\exp\left\{-\frac{b^2}{2\sigma_Y^2}\right\} \tag{6.9}$$

According to the criterion of the minimum failure rate, the mean level-crossing rate should be reduced as much as possible in the structural design in order to achieve higher time-variant reliability of structures against random vibration. It is observed from Eq. (6.9) that the mean level-crossing rate could be effectively decreased by reducing the RMS value of the response process $Y(t)$.

Combining Eq. (6.2) and Eq. (6.9), the Poisson model based estimates of the probability of no crossing for double barriers are

$$R(t) = P_0(t) = \exp\left[-\frac{1}{\pi}\frac{\sigma_{\dot{Y}}}{\sigma_Y}\exp\left(-\frac{b^2}{2\sigma_Y^2}\right)t\right] \tag{6.10}$$

Though convenient, the Poisson model proved to be too conservative when the response process $Y(t)$ is narrowband and/or when the threshold $b$ is not high enough with respect to the RMS value of the response (Vanmarcke 1975). In these situation, consecutive down- or up-crossings of the absolute response value $|Y(t)|$, cannot be realistically assumed as independent events, as they tend to occur in clumps. Vanmarcke (1975) developed an improved formula of mean level-crossing rate based on a modified Poisson model, accounting for the dependence between the crossing events. The corrected mean out-crossing rate are given as

$$\eta_b = \nu_b \frac{1-\exp\left(-\sqrt{\frac{\pi}{2}}q^{1.2}\frac{b}{\sigma_Y}\right)}{1-\exp\left(-\frac{b^2}{2\sigma_Y^2}\right)} \tag{6.11}$$

where $q = \sqrt{1-\lambda_1^2/(\lambda_0\lambda_2)}$ =shape factor that characterizes the bandwidth of the process, in

144

which $\lambda_m$ =spectral moments defined by

$$\lambda_m = \int_0^\infty \omega^m G_Y(\omega)d\omega, \; m = 0, 1, 2, 4 \tag{6.12}$$

where $G_Y(\omega)$ =one-sided power spectral density function of the process. Obviously, one has $\lambda_0 = \sigma_Y^2$, $\lambda_2 = \sigma_{\dot{Y}}^2$ and $\lambda_4 = \sigma_{\ddot{Y}}^2$. It is also worth noting that in Eq. (6.11) only the exponent 1.2 of $q$ is of empirical nature, and it was recommended by Vanmarcke in order to improve the agreement with the results of Monte Carlo simulations for the stationary response of a lightly damped linear oscillator subjected to Gaussian white-noise.

### 6.2.1.2 Stationary combined random process

Under wind load excitation, a building may move in a lateral-torsional manner such that the maximum resultant response may involve with several component responses. Assuming that the corner of a building experiences two perpendicular translational component responses, $X(t)$ and $Y(t)$. Then the combined resultant process can be written as

$$A(t) = \sqrt{X^2(t) + Y^2(t)} \tag{6.13}$$

Assuming $X(t)$ and $Y(t)$ following Gaussian distribution with zero-mean and common standard deviation of $\sigma_X = \sigma_Y$, then $A(t)$ follows Rayleigh distribution (Cramer 1946). The Rayleigh distribution of $A(t)$ is given as

$$f_A(a) = \frac{a}{\sigma_A^2} \exp\left(-\frac{a^2}{2\sigma_A^2}\right) \tag{6.14}$$

where $\sigma_A$ denotes the mode value of $A(t)$. Since it is assumed that the two component processes have the common standard deviation $\sigma_X = \sigma_Y$, the mode value of $A(t)$ coincides with the component standard deviation, i.e., we have $\sigma_A = \sigma_X = \sigma_Y$.

Taking derivative at both sides of Eq. (6.13) with respect to $t$, gives

$$\dot{A}(t) = \frac{X(t)}{A(t)}\dot{X}(t) + \frac{Y(t)}{A(t)}\dot{Y}(t) \tag{6.15}$$

Since the derivative process $\dot{A}(t)$ can be approximated by a linear combination of $\dot{X}(t)$ and $\dot{Y}(t)$ as shown in Eq. (6.15), the probability density function $f_{\dot{A}}(\dot{a})$ of the time derivative process $\dot{A}(t)$ of

145

the Rayleigh process $A(t)$ is approximately Gaussian and can be given by

$$f_{\dot{A}}(\dot{a}) = \frac{a}{\sqrt{2\pi}\sigma_{\dot{A}}} \exp\left(-\frac{\dot{a}^2}{2\sigma_{\dot{A}}^2}\right) \tag{6.16}$$

where $\sigma_{\dot{A}}$ indicates the standard deviation value of the derivative process $\dot{A}(t)$. It is reasonable to assume that the process $A(t)$ is independent of its derivative process $\dot{A}(t)$. Therefore, the joint probability density function given in the Rice formula for calculating the mean level-crossing rate can be determined as

$$f_{A,\dot{A}}(a,\dot{a}) = f_A(a)f_{\dot{A}}(\dot{a}) = \frac{a}{\sqrt{2\pi}\sigma_A^2\sigma_{\dot{A}}} \exp\left[-\frac{1}{2}\left(\frac{a^2}{\sigma_A^2} + \frac{\dot{a}^2}{\sigma_{\dot{A}}^2}\right)\right] \tag{6.17}$$

Substituting Eq. (6.17) into Eq. (6.7), the mean upcrossing rate of level b for the Rayleigh process $A(t)$ is obtained as

$$v_b^+ = \int_0^\infty \dot{a} f_{A,\dot{A}}(b,\dot{a})d\dot{a} = \frac{b\sigma_{\dot{A}}}{\sqrt{2\pi}\sigma_A^2} \exp\left(-\frac{b^2}{2\sigma_A^2}\right) \tag{6.18}$$

Substituting Eq. (6.18) into (6.2), the Poisson model based estimate of the probability of no upcrossing from single barrier $b$ for the combined resultant process $A(t)$ is

$$R_A(t) = \exp\left[-\frac{b\sigma_{\dot{A}}}{\sqrt{2\pi}\sigma_A^2} \exp\left(-\frac{b^2}{2\sigma_A^2}\right)t\right] \tag{6.19}$$

## 6.2.2 Extreme value distribution and probabilistic peak factor

The largest absolute peak response over a given time duration $\tau$ can be defined as a new random variable

$$Y_\tau = \max\left\{|Y(t)|; 0 \leq t \leq \tau\right\} \tag{6.20}$$

And let the first-passage time $T_b$ be a random variable describing the instant at which the structural absolute response $|Y(t)|$ firstly crosses the barrier of level $b$. The largest absolute response does not exceed the level $b$ if and only if the corresponding first-passage time $T_b$ does not occur prior to $\tau$. Thus, the equivalence of the events $Y_\tau \leq b$ and $T_b \geq \tau$ yields the fundamental relationship between the distributions of the largest value and the first-passage time (also the first-passage

146

probability)

$$F_{Y_\tau} = P[Y_\tau \le b] = P[T_b \ge \tau] = R(\tau) \tag{6.21}$$

From Eq. (6.10), the cumulative probability distribution function of $Y_\tau$ can be expressed as

$$F_{Y_\tau}(b) = \exp\left[-v_0\tau\exp\left(-\frac{b^2}{2\sigma_Y^2}\right)\right] \tag{6.22}$$

where $v_0 = \dfrac{1}{\pi}\dfrac{\sigma_{\dot{Y}}}{\sigma_Y}$, is the mean zero-crossing rate of the basic process $Y(t)$. Let $b_{\tau,p}$ denote the

values of $Y_\tau$, the largest peak response of $|Y(t)|$ during $\tau$ seconds of stationary motion, which has a prescribed probability, $p$, of not being exceeded. It may be obtained by solving the following equation

$$p = F_{Y_\tau}(b_{\tau,p}) = \exp\left[-v_0\tau\exp\left(-\frac{b_{\tau,p}^2}{2\sigma_Y^2}\right)\right] \tag{6.23}$$

By taking logarithm twice at both sides of Eq. (6.23) successively, yields

$$\frac{b_{\tau,p}}{\sigma_Y} = \sqrt{2\ln\frac{v_0\tau}{\ln(1/p)}} = g_e \tag{6.24}$$

The ratio of $b_{\tau,p}/\sigma_Y$ is also be known as the peak factor. Eq. (6.24) can be regarded as an approximation of the peak factor, and called as the equivalent peak factor ($g_e$) in this study for comparison purpose.

Davenport (1964) has shown that, if the underlying marginal distribution of a process is Gaussian, then the largest values of the process will asymptotically follow a Gumbel distribution. Davenport's peak factor is given as

$$g_f = \sqrt{2\ln v_0\tau} + \gamma/\sqrt{2\ln v_0\tau} \tag{6.25}$$

where $\gamma = 0.5772$, is the Euler's constant. In wind engineering practice, $v_0$ could be approximated by the natural frequency of a building, while observation time duration may be taken as 3600s. It is worth mentioning that the above expression is used to calculate the expected largest or maximum peak response. The Davenport's peak factor is independent of bandwidth parameter, and approximately related to the exceedance probability of 50% (Davenport 1964). Once Eq. (6.25)

147

being used to estimate the expected maximum response, the probability of the expected maximum response not being exceeded could be evaluated by the following equation based on the Gumbel distribution (Type I asymptotic distribution of extreme value) as

$$P(Y_\tau \le g_f \sigma_R) = \exp\left[-\exp\left(-g_f \sqrt{2\ln v_0 \tau} + 2\ln v_0 \tau\right)\right] = e^{-e^{-\gamma}} = 0.57 \quad (6.26)$$

Vanmarcke's formula of Eq. (6.11) for the corrected mean out-crossing rate of level $b$ could be used in Eq. (6.2) as a replacement of $v_b$. Therefore, the cumulative probability distribution function of $Y_\tau$ may be rewritten as

$$F_{Y_\tau}(b) = \exp\left[-v_0 \tau \frac{1 - \exp\left(-\sqrt{\frac{\pi}{2}} q^{1.2} \frac{b}{\sigma_Y}\right)}{\exp\left(\frac{b^2}{2\sigma_Y^2}\right) - 1}\right] \quad (6.27)$$

Starting from Eq. (6.27), it is possible to derive an explicit expression of peak factor, called as probabilistic peak factor, which depends on the specified value of the probability of no exceedance ($p$), but also on the shape parameter of PSD of the random response process ($q$) and the excitation duration time ($\tau$). Denoting $b_{\tau,p} / \sigma_Y = g_p(p, q, \tau)$ and substituting $g_p$ and $p$ into Eq. (6.27), gives

$$p = \exp\left[-v_0 \tau \frac{1 - \exp\left(-\sqrt{\frac{\pi}{2}} q^{1.2} g_p\right)}{\exp\left(\frac{g_p^2}{2}\right) - 1}\right] \quad (6.28)$$

Taking logarithm at both sides of (6.28), and rearranging terms, one obtains

$$\frac{v_0 \tau}{\ln(1/p)}\left[1 - \exp\left(-\sqrt{\frac{\pi}{2}} q^{1.2} g_p\right)\right] = \exp\left(\frac{g_p^2}{2}\right) - 1 \quad (6.29)$$

It is noted that $\exp\left(g_p^2/2\right) \gg 1$ and $\exp\left(-\sqrt{\pi/2} q^{1.2} g_p\right) \approx \exp\left(-\sqrt{\pi/2} q^{1.2} g_e\right)$, then Eq. (6.29) could be written approximately as

$$\frac{v_0 \tau}{\ln(1/p)}\left[1 - \exp\left(-\sqrt{\frac{\pi}{2}} q^{1.2} g_e\right)\right] \approx \exp\left(\frac{g_p^2}{2}\right) \quad (6.30)$$

Hence, the following semi-empirical equation for the probabilistic peak factor may be used to represent the solution of Eq. (6.28)

$$g_p = \sqrt{2\ln\left\{\frac{v_0\tau}{\ln(1/p)}\left[1-\exp\left(-\sqrt{\pi/2}q^{1.2}g_e\right)\right]\right\}}$$   (6.31)

The peak factors, calculated according to the equivalent peak factor of Eq. (6.24), Davenport's peak factor of Eq. (6.25) and the probabilistic peak factor of Eq. (6.31), are plotted in Figure 6.1 for several values of $q$ and for $v_0 = 0.1 \sim 1.1$Hz, which represents the range of frequency values for typical tall buildings. For keeping consistent with Davenport's peak factor, the probability of no exceedance $p$ is taken as 0.57 and the excitation duration time $\tau$ is 3600 s. It is found that the equivalent peak factor agrees almost exactly the same as the Davenport's peak factor. Both of two factors are independent of bandwidth parameter $q$, and tend to give conservative results. From this figure, the influence of the bandwidth parameter $q$ is found to be significant on the value of probabilistic peak factor. When $q$ approaches to 1, the probabilistic peak factor also approaches to the value of the upper bound curve for the Davenport's peak factor as well as the equivalent peak factor.

## 6.3   Asymptotic theory of statistical extremes

### 6.3.1   Analytical peak distribution for stationary random process

#### 6.3.1.1   Peak distribution of stationary Gaussian random process

The analytical peak distribution for a stationary normal random process was firstly studied by Rice (1945). Based on the Rayleigh distribution of peaks, Davenport's peak factor can be obtained using the asymptotic theory of statistical extremes. More specific information of peak distribution may lead to more accurate estimation of the expected largest peak response, or simply speaking, the expected maximum response. This fact motivates the further study of the probability distribution of peaks. For a data sample of random response process, a peak occurs at the time instant $t$ under the condition that $\dot{y}(t) = 0$ and $\ddot{y}(t) < 0$. Similarly, a trough occurs at the condition that $\dot{y}(t) = 0$ and $\ddot{y}(t) > 0$. Hence the zero-one process may be constructed in order to count the number of peaks or troughs as

$$Z_m(t) = u[\dot{Y}(t)]$$   (6.32)

where the subscript $m$ indicates the local maximum values attained at peaks or troughs. The derivative of $Z_m(t)$ yields the number of extrema per second as the desired counting function

$$\dot{Z}_m(t) = \ddot{Y}(t)\delta[\dot{Y}(t)] \qquad (6.33)$$

Actually, $\delta[\dot{Y}(t)]$ gives a spike (at time $t$) whenever $\dot{y}(t) = 0$. The total number of extrema (local maxima, including peaks and troughs) in the time interval $(t_2\text{-}t_1)$ is

$$N_m(t_1,t_2) = \int_{t_1}^{t_2} |\ddot{Y}(t)| \delta\left[\dot{Y}(t)\right] dt \qquad (6.34)$$

The unit step function $u[Y(t) - b]$ can be introduced into the above equation in order to ensure that the extrema counting operation is carried out for all $Y(t) \geq b$. It is noted that such a consideration is consistent with the peak over threshold (POT) approach for the extreme value regression analysis. Then, the total number of extrema for $Y(t) \geq b$ in the interval $(t_2\text{-}t_1)$ is

$$N_{m,b}(t_1,t_2) = \int_{t_1}^{t_2} |\ddot{Y}(t)| \delta\left[\dot{Y}(t)\right] u[Y(t) - b] dt \qquad (6.35)$$

The expected number of peaks per second of $y$ above the level $b$ for all $\ddot{y} < 0$, becomes

$$
\begin{aligned}
V_b^+ &= E\left[N_{m,b}(t_1,t_2)\right] / dt = E\left\{|\ddot{Y}(t)| \delta\left[\dot{Y}(t)\right] u[Y(t) - b]\right\} \\
&= \int_{-\infty}^{\infty} \int_{-\infty}^{\infty} \int_{-\infty}^{\infty} |\ddot{y}| \delta[\dot{y}] u[y - b] f_{Y,\dot{Y},\ddot{Y}}(y,\dot{y},\ddot{y}) dy d\dot{y} d\ddot{y} \qquad (6.36) \\
&= -\int_{-\infty}^{0} d\ddot{y} \int_{b}^{\infty} \ddot{y} f_{Y,\dot{Y},\ddot{Y}}(b,0,\ddot{y}) dy
\end{aligned}
$$

where the superscript + denotes the upward crossing; $f_{Y,\dot{Y},\ddot{Y}}(y,\dot{y},\ddot{y})$ is the joint probability density function of $Y(t)$, its derivative $\dot{Y}(t)$ and its second derivative $\ddot{Y}(t)$. Assuming that $Y(t)$ is a Gaussian stationary process with zero mean, the joint probability density function can be given as

$$f_{Y,\dot{Y},\ddot{Y}}(y,0,\ddot{y}) = \frac{1}{(2\pi)^{3/2}|\Lambda|} \exp\left[-\frac{1}{2|\Lambda|}\left(\sigma_{\dot{Y}}^2 \sigma_{\ddot{Y}}^2 y^2 + 2\sigma_{\dot{Y}}^4 y\ddot{y} + \sigma_Y^2 \sigma_{\dot{Y}}^2 \ddot{y}^2\right)\right] \qquad (6.37)$$

where the determinant of covariance matrix is calculated by

$$|\Lambda| = \begin{vmatrix} \sigma_Y^2 & 0 & -\sigma_{\dot{Y}}^2 \\ 0 & \sigma_{\dot{Y}}^2 & 0 \\ -\sigma_{\dot{Y}}^2 & 0 & \sigma_{\ddot{Y}}^2 \end{vmatrix} \qquad (6.38)$$

in which $\sigma_{\ddot{Y}}$ = the RMS value of the derived process $\ddot{Y}(t)$.

Based on the frequentist definition of probability, the cumulative probability distribution function for peaks can be related to the ratio of the expected number of peaks below the level $b$ per second over the expected number of total peaks per second as

$$F_{Y_m}(b) = \frac{V_{-\infty}^+ - V_b^+}{V_{-\infty}^+} \quad (6.39)$$

where $V_{-\infty}^+$ denotes the expected number of total peaks per second. Substituting Eq. (6.36) into Eq. (6.39) and differentiating the resultant equation with respect to $b$ gives the PDF of peaks

$$f_{Y_m}(b) = -\frac{1}{V_{-\infty}^+} \int_{-\infty}^0 \ddot{y} f_{Y,\dot{Y},\ddot{Y}}(b,0,\ddot{y}) d\ddot{y} \quad (6.40)$$

As a normalizing constant, $V_{-\infty}^+$ could be readily obtained from Eq. (6.9). The occurrence of a peak (i.e., when $\dot{y}(t) = 0$ and $\ddot{y}(t) < 0$) means that the process $\dot{Y}(t)$ crosses zero with a negative slop. Consider $\dot{Y}(t)$ as the basic process, the mean zero-crossing rate of $\dot{Y}(t)$ is equal to the average rate of occurrence of local maxima (including peaks and troughs)

$$v_{0,\dot{Y}} = \frac{1}{\pi} \frac{\sigma_{\ddot{Y}}}{\sigma_{\dot{Y}}} = 2V_{-\infty}^+ \quad (6.41)$$

Substituting Eq. (6.37) and Eq. (6.41) into Eq. (6.40) and integrating over $\ddot{y}$, one obtains

$$\begin{aligned}
f_{Y_m}(b) = &\frac{\varepsilon}{\sqrt{2\pi}\sigma_Y} \exp\left(-\frac{b^2}{2\varepsilon^2\sigma_Y^2}\right) \\
&+ \frac{b\sqrt{1-\varepsilon^2}}{\sigma_Y^2} \exp\left(-\frac{b^2}{2\sigma_Y^2}\right)\left[\frac{1}{2} + \frac{1}{2}\mathrm{erf}\left(\sqrt{\frac{1-\varepsilon^2}{2}}\frac{b}{\sigma_Y\varepsilon}\right)\right]
\end{aligned} \quad (6.42)$$

where $\varepsilon = \sqrt{1 - \lambda_2^2/(\lambda_0\lambda_4)}$ = dimensionless spectral parameter that serves as a measure of bandwidth, but is less direct than $q$ due to the use of higher-order moments; the error function $\mathrm{erf}(z) = \frac{2}{\sqrt{\pi}} \int_0^z \exp(-u^2) du$. For a pure narrowband process, in which $\varepsilon = 0$, the first term of Eq. (6.42) vanishes and the equation is then reduced to the well-known Rayleigh distribution as

$$f_{Y_m}(b) = \frac{b}{\sigma_Y^2} \exp\left(-\frac{b^2}{2\sigma_Y^2}\right) \quad (6.43)$$

Since the mode value (at which, the probability density function attains its maximum value) of the

151

Rayleigh distribution is $\sigma_Y$, the majority of the peaks approaches to the value of $\sigma_Y$ for a narrowband process. When $\varepsilon = 1$, Eq. (6.42) is reduced to the Gaussian PDF with a zero mean and a variance of $\sigma_Y^2$. This implies that local peaks tend to occur erratically, with equal chance of being below and above the mean of $Y(t)$.

In reality, a random response process is neither pure narrow band nor pure wideband, and has a bandwidth parameter of $0 < \varepsilon < 1$. Hence, even for a random response process with Gaussian distribution, the PDF of peaks follows the distribution given in Eq. (6.42), which is neither the Rayleigh nor the Gaussian distribution due to $0 < \varepsilon < 1$. Figure 6.2 shows the PDF curves of peaks corresponding to various value of $\varepsilon$ with $\sigma_Y = 1$. It is evidenced shown in Figure 6.2 that the peak distribution of a Gaussian process asymptotically approaches to the Rayleigh distribution as $\varepsilon$ gets close to zero.

### 6.3.1.2  Peak distribution of stationary Rayleigh random process

The combined random process $A(t)$ defined in Eq. (6.13) generally follows non-Gaussian distribution. If two component processes have the common standard deviation, $A(t)$ follows Rayleigh distribution as given in Eq. (6.14). While the peak distribution of a Gaussian process has been discussed in the last section, the peak distribution of a non-Gaussian response process, particularly of a combined response process $A(t)$, is studied in this section. The cumulative probability distribution function of peaks for a stationary random process is given in Eq. (6.39), which is applicable to any stationary process either Gaussian or non-Gaussian. Hence, the PDF of peaks for a stationary combined process $A(t)$ can be obtained by taking derivative with respect to $b$ at both sides of Eq. (6.39) as

$$f_{A_m}(b) = -\frac{1}{V_{-\infty}^+} \frac{dV_b^+}{db} \tag{6.44}$$

where $A_m$ denotes a random variable, defined as peak values arising from a combined process $A(t)$. For simplification, it is assumed in this section that the combined response process $A(t)$ follows Rayleigh distribution.

In theory, the expected number of peaks per second $V_b^+$ can be calculated by Eq. (6.36). The joint probability density function, however, is generally not available for a non-Gaussian process. For a narrow-band process, each upcrossing event can possibly lead to find a corresponding peak. The expected number of peaks above the given level $b$ per second can then be approximated by the level upcrossing rate $v_b^+$. The expected number of total peaks per second $V_{-\infty}^+$ can also be assumed to be equal to the mean upcrossing rate from the relatively lower level of mode value $\sigma_A$, i.e., $V_{-\infty}^+ \approx v_b^+ \big|_{b=\sigma_A}$. Eq. (6.44) can then be rewritten as

$$f_{A_m}(b) \approx -\frac{1}{v_{\sigma_A}^+} \frac{dv_b^+}{db} \tag{6.45}$$

Substituting Eq. (6.18) into Eq. (6.45), the peak distribution in terms of PDF for a Rayleigh process can be obtained analytically as

$$f_{A_m}(b) \approx \frac{1}{\sigma_A}\left(\frac{b^2}{\sigma_A^2}-1\right)\exp\left[-\frac{1}{2}\left(\frac{b^2}{\sigma_A^2}-1\right)\right], \quad b \geq \sigma_A \tag{6.46}$$

By introducing the intermediate threshold level $c = b^2/\sigma_A^2 - 1$ corresponding to a peak-dependent intermediate random variable $C = A_m^2/\sigma_A^2 - 1$, which is called as the intermediate peak variable, the elementary probability of the event for an occurring peak with $\{b \leq A_m \leq b+db\}$ can be related to the elementary probability of the event for the intermediate peak variable with $\{c \leq A_m^2/\sigma_A^2 - 1 \leq c+dc\}$ as

$$f_{A_m}(b)db = \frac{c}{2\sqrt{c+1}}\exp\left[-\frac{1}{2}c\right]dc, \quad c \geq 0 \tag{6.47}$$

Therefore, the distribution behavior of the intermediate peak variable can be described by the PDF

$$f_C(c) = \frac{c}{2\sqrt{c+1}}\exp\left[-\frac{1}{2}c\right], \quad c \geq 0 \tag{6.48}$$

It is observed that the above probability model is similar to one simple form of Gamma distributions, which is given by

$$f_C(c) \approx \frac{c}{4}\exp\left[-\frac{1}{2}c\right], \quad c \geq 0 \tag{6.49}$$

For wind related time-variant reliability problems, a desired threshold level $b$ is normally larger than $3\sigma_A$ such that the intermediate threshold $c$ is larger than 8 as $c = b^2/\sigma_A^2 - 1$. In this situation, the Gamma probability model may generally result in higher value of PDF due to the fact

153

that $\frac{c}{4} > \frac{c}{2\sqrt{c+1}}$. The higher value in the tail range of the probability density function indicates the Gamma probability model is more conservative in its estimation of the crossing failure probability. Figure 6.3 shows the PDF curves of peaks and the intermediate peak variable for a Rayleigh process with $\sigma_A = 1$. It is noted that the tail distribution behavior of the intermediate peak variable can be fairly approximated by the Gamma distribution.

## 6.3.2 Asymptotic extreme value distribution and expected peak factor

### 6.3.2.1 Davenport's peak factor and the Weibull peak factor

From a viewpoint of the statistics of extremes, the largest peak values, i.e., extreme values, can be statistically obtained from a sample of peak response values with size $n$ as

$$Y_n = \max\left(Y_{m_1}, Y_{m_2}, ..., Y_{m_n}\right) \tag{6.50}$$

where $Y_{m_1}, Y_{m_2}, ..., Y_{m_n}$ are assumed to be statically independent and identically distributed with the same peak distribution. If the peak distribution of interest follows the Rayleigh distribution, the distribution of extreme values from the initial variate of peaks will converge to the Extreme Value Type I (Gumbel) distribution (Gumbel 1958) for a sufficiently large value of $n$. Given the Rayleigh PDF of Eq. (6.43), the cumulative distribution function of $Y_m$ is

$$F_{Y_m}(b) = 1 - \exp\left(-\frac{b^2}{2\sigma_Y^2}\right) \tag{6.51}$$

The cumulative distribution function of Extreme Value Type I for the distribution of the largest value is given as follows

$$F_{Y_n}(b) = \exp\left[-e^{-(b-u_n)/\beta_n}\right] \tag{6.52}$$

where $u_n$ and $\beta_n$ are, respectively, the location and scale parameters. The characteristic largest value, $u_n$, is defined as the particular value of $Y_m$ such that in a sample of size $n$ from the initial population $Y_m$, the expected number of sample values larger than $u_n$ is one; that is

$$F_{Y_m}(u_n) = 1 - \frac{1}{n} \tag{6.53}$$

154

Therefore, according to Eq. (6.51),

$$1 - \exp\left(-\frac{u_n^2}{2\sigma_Y^2}\right) = 1 - \frac{1}{n} \tag{6.54}$$

From which, one obtains

$$u_n = \sigma_Y \sqrt{2 \ln n} \tag{6.55}$$

The scale parameters, $\beta_n$, is a measure of dispersion of $Y_n$, can be determined by

$$\beta_n = \left[ n f_{Y_m}(u_n) \right]^{-1} = \left[ n \frac{u_n}{\sigma_Y^2} \exp\left(-\frac{u_n^2}{2\sigma_Y^2}\right) \right]^{-1} = \frac{\sigma_Y}{\sqrt{2 \ln n}} \tag{6.56}$$

The mean $\mu_{Y_n}$ and standard deviation $\sigma_{Y_n}$ of the extreme value $Y_n$, therefore, are obtained as follows

$$\mu_{Y_n} = u_n + \gamma \beta_n = \sigma_Y \left( \sqrt{2 \ln n} + \gamma / \sqrt{2 \ln n} \right) \tag{6.57}$$

$$\sigma_{Y_n} = \frac{\pi \sigma_Y}{2\sqrt{3 \ln n}} \tag{6.58}$$

where $\gamma$ is the Euler's constant. For a narrowband response process, one peak occurs possibly accompanied with one zero-crossing event. Therefore, the sample size of peaks as well as troughs, $n$, may be determined from the mean zero-crossing rate $v_0$ of the response process. Hence, for a given time duration $\tau$, one has $n = v_0 \tau$. From Eq. (6.57), the expected peak factor could be expressed as

$$\frac{\mu_{Y_n}}{\sigma_Y} = \sqrt{2 \ln v_0 \tau} + \gamma / \sqrt{2 \ln v_0 \tau} \tag{6.59}$$

which is exactly the same as the Davenport's peak factor given in Eq. (6.25).

However, as discussed in the previous section, the real distribution of peaks may neither be Rayleigh nor Gaussian. As such, the real peak distribution is a compromise between the Rayleigh and Guassian, depending on the spectral parameter of the PSD of the basic process $Y(t)$. Based on Eq. (6.42), it is difficult to obtain a general closed-form expression of the expected peak factor for $0 < \varepsilon < 1$. As a reference, Cramer (1946) gives the following asymptotic solution for the expected peak factor of $Y_n$ with the assumption of a Gaussian distribution of peaks

$$\frac{\mu_{Y_n}}{\sigma_Y} = \sqrt{2\ln n} + \frac{\gamma}{\sqrt{2\ln n}} - \frac{\ln\ln n + \ln 4\pi}{2\sqrt{2\ln n}} \tag{6.60}$$

When the actual distribution of peaks departs significantly from a Rayleigh distribution or a Gaussian distribution, the more general Weibull distribution can be used to model the distribution of peaks (Melbourne 1977; Newland 1984; Cheng et al. 2003). The Weibull distribution is given as

$$f_{Y_m}(b) = \frac{\kappa}{\rho}\left(\frac{b^{\kappa-1}}{\sigma_Y^{\kappa}}\right)\exp\left[-\frac{1}{\rho}\left(\frac{b}{\sigma_Y}\right)^{\kappa}\right] \quad \text{for } b > 0 \tag{6.61}$$

where $\kappa$ denotes the shape parameter of the peak distribution; $\rho$ indicates the scale parameter, and is determined by $\rho = \left(\sqrt{2\ln 2}\right)^{\kappa}/\ln 2$ (Newland 1984). In case of $(\kappa, \rho) = (2, 2)$, the Weibull distribution is reduced into the Rayleigh distribution with variance $\sigma_Y^2$. This degeneration indicates that the Weibull distribution can be regarded as a generalization of the Rayleigh distribution.

Similar to the case of the Rayleigh distribution of peaks, a closed-form expression of the expected peak factor could also be derived from the Weibull distribution. Given the Weibull distribution as the initial peak variate, the corresponding cumulative distribution function of peaks, $Y_m$, can be given as

$$F_{Y_m}(b) = P(Y_m \le b) = 1 - \exp\left[-\frac{1}{\rho}\left(\frac{b}{\sigma_Y}\right)^{\kappa}\right] \tag{6.62}$$

For the corresponding Extreme Value Type I distribution, the characteristic largest value $u_n$ with the probability of exceedance defined as $1/n$ can be determined by the following equation

$$F_{Y_m}(u_n) = 1 - \exp\left[-\frac{1}{\rho}\left(\frac{u_n}{\sigma_Y}\right)^{\kappa}\right] = 1 - \frac{1}{n} \tag{6.63}$$

From which, one obtains the characteristic largest value as

$$u_n = \sigma_Y\left(\rho\ln n\right)^{1/\kappa} \tag{6.64}$$

Thus the scale parameter, $\beta_n$, can be determined by

$$\beta_n = \left[ n f_{Y_m}(u_n) \right]^{-1} = \left[ n \frac{\kappa}{\rho} \left( \frac{u_n^{\kappa-1}}{\sigma_Y^\kappa} \right) \exp\left[ -\frac{1}{\rho} \left( \frac{u_n}{\sigma_Y} \right)^\kappa \right] \right]^{-1} = \frac{\sigma_Y (\rho \ln n)^{1/\kappa}}{\kappa \ln n} \quad (6.65)$$

The expected peak factor for the Weibull peak distribution can be written as

$$g_W = \frac{\mu_{Y_n}}{\sigma_Y} = \frac{u_n + \gamma \beta_n}{\sigma_Y} = (\rho \ln n)^{1/\kappa} + \frac{\gamma (\rho \ln n)^{1/k}}{\kappa \ln n} \quad (6.66)$$

Let $n = v_0 \tau$ and $(\kappa, \rho) = (2, 2)$, the above equation is reduced into Eq. (6.59), which is the expected peak factor for the Rayleigh distribution. For the sake of comparison, the peak factor $g_W$ given by Eq. (6.66) would be called as the Weibull peak factor.

The shape parameter of the Weibull distribution can be estimated by statistically analyzing the sample records of a random response process of interest using the probability paper method (Gumbel 1958; Melbourne 1977; Newland 1984). Based on the data samples of an acceleration response, the peak acceleration distribution and the largest peak acceleration distribution can be statistically estimated. Conventionally, only one single global maximum value of each simulated data sample is used in the statistical analysis of the largest peak response. For accuracy reason, a large number of samples has to be prepared by repeated experimental tests, or a computer-based simulation or analysis is used to estimate the maximum response (Holmes and Cochran 2003; Cheng et al. 2003). By restricting the data to the global maximum of each simulation, only a small portion of the potentially useful information can be utilized in the maximum value approach. This drawback can be avoided by using the POT (peak over threshold) approach to estimate peak distribution, in which more information in the analysis can be incorporated by including local maxima above a certain chosen threshold in the statistical analysis. The threshold value can be typically taken as the standard deviation of the acceleration response. The selected data of peaks are then fitted with the Weibull distribution using the probability paper method (Gumbel 1958). From Eq. (6.62), one obtain

$$P(Y_m > b) = \exp\left[ -\frac{1}{\rho} \left( \frac{b}{\sigma_Y} \right)^\kappa \right] \quad (6.67)$$

Substituting $\rho = \left( \sqrt{2 \ln 2} \right)^\kappa / \ln 2$ into Eq. (6.67), and taking logarithm twice on the resulting equation, the above equation becomes

$$\ln\left\{-\ln\left[P\left(Y_m > b\right)\right]\right\} = \ln\ln 2 + \kappa \ln\left(\frac{b}{\sigma_Y}\right) - \kappa \ln\sqrt{2\ln 2} \qquad (6.68)$$

Hence the slope of a graph of $\ln\left\{-\ln\left[P\left(Y_m > b\right)\right]\right\}$ against $\ln\left(\dfrac{b}{\sigma_Y}\right)$ is the Weibull exponent $\kappa$.

### 6.3.2.2 Gamma peak factor for a combined response process

It is also possible to obtain a closed-form formula of peak factors for a combined resultant response process $A(t) = \sqrt{X^2(t) + Y^2(t)}$ as discussed in section 6.2.1.2. Assuming $X(t)$ and $Y(t)$ following Gaussian distribution with zero-mean and common standard deviation of $\sigma_X = \sigma_Y$, then $A(t)$ follows Rayleigh distribution. The Gamma distribution in Eq. (6.49) for the intermediate peak variable $C = A_m^2/\sigma_A^2 - 1$ has a corresponding analytical cumulative distribution function of $C$ as following

$$F_C(c) = P\left(C \le c\right) = 1 - \left(1 + \frac{1}{2}c\right)\exp\left(-\frac{1}{2}c\right), \quad c \ge 0 \qquad (6.69)$$

It can be shown that the largest value arising from the Gamma distribution will converges to the Type I asymptotic form (Gumbel, 1958), i.e., Eq. (6.52). The $u_n$ and $\beta_n$ of the corresponding Extreme Value Type I distribution for the reduced Gamma distribution can be determined by the following two equations

$$\left(1 + \frac{1}{2}u_n\right)\exp\left[-\frac{1}{2}u_n\right] = \frac{1}{n} \qquad (6.70)$$

$$\beta_n = \left[nf_C(u_n)\right]^{-1} = \left[n\frac{u_n}{4}\exp\left(-\frac{1}{2}u_n\right)\right]^{-1} \qquad (6.71)$$

Taking logarithm at both sides of Eq. (6.70), gives

$$u_n = 2\ln n + 2\ln(1 + 0.5u_n) \qquad (6.72)$$

If $n$ is sufficiently large, (e.g., $n>100$), the characteristic largest value for the Gamma distribution can thus be approximated as

$$u_n \approx 2\ln n + 2\ln\ln n \qquad (6.73)$$

The dispersion of the largest value from the reduced Gamma distribution is also obtained using Eq.

(6.71) as follows

$$\beta_n \approx \frac{2\sqrt{\ln n}}{\ln n + \ln \ln n} \tag{6.74}$$

The mean and standard deviation of the largest value for the intermediate peak variable are given as follows

$$\mu_{C_n} = u_n + \gamma\beta_n = 2\ln n + 2\ln \ln n + \frac{2\gamma\sqrt{\ln n}}{\ln n + \ln \ln n} \tag{6.75}$$

$$\sigma_{C_n} = \frac{\pi\sqrt{2\ln n}}{\sqrt{3}\left(\ln n + \ln \ln n\right)} \tag{6.76}$$

Using the relationship of the intermediate peak variable to the peaks of a Rayleigh process, i.e., $C = A_m^2 / \sigma_A^2 - 1$, the expected peak factor of a Rayleign resultant process can be written as

$$g_G = \sqrt{\mu_{C_n} + 1} = \left(2\ln n + 2\ln \ln n + \frac{2\gamma\sqrt{2\ln n}}{\ln n + \ln \ln n} + 1\right)^{1/2} \tag{6.77}$$

The above equation gives a closed-form formula to estimate the expected largest peak resultant response for a combined resultant process with the Rayleigh distribution. For comparison, the peak factor determined by Eq. (6.77) is called as the Gamma peak factor. The expected maximum resultant response can then be estimated in terms of the Gamma peak factor and the mode value of the resultant process $A(t)$ as

$$\mu_{A_n} = g_G\sigma_A = g_G\sigma_Y \tag{6.78}$$

As discussed in Chapter 5, the standard deviation of component acceleration response of a wind-excited tall building can be readily estimated by spectral analysis in the frequency domain. Eq. (6.78) provides a more accurate closed-form formula, compared to Eq. (5.26) or Eq. (5.27), to estimate the expected maximum resultant response of a wind-sensitive tall building. Unlike the statistic Weibull peak factor expressed by Eq. (6.66), the evaluation of the Gamma peak factor in Eq. (6.77) and in turn the use of Eq. (6.78) do not require any time-history response data.

Eq. (6.78) is only applicable to a resultant process with two components having the same standard deviation. If two components do not have the same value of standard deviation, i.e., $\sigma_X \neq \sigma_Y$, the combined resultant process $A(t)$ will have a more complicated probability distribution than the Rayleigh distribution. Assuming that $X(t)$ and $Y(t)$ can be described by the zero-mean Gaussian

159

distribution $N(0, \sigma_X)$ and $N(0, \sigma_Y)$, a reduced resultant process can be defined as

$$\tilde{A}(t) = \sqrt{\left(\frac{X(t)}{\sigma_X}\right)^2 + \left(\frac{Y(t)}{\sigma_Y}\right)^2}$$

(6.79)

It can be shown that the probability model of $\tilde{A}(t)$ can be described by a Rayleigh distribution with a mode value of unity. Then the expected largest response of $\tilde{A}(t)$ can be readily obtained by Eq. (6.78) as

$$\mu_{\tilde{A}_n} = g_G$$

(6.80)

Assuming $\sigma_Y > \sigma_X$, the resultant process can be rewritten in terms of $Y(t)$ and $\tilde{A}(t)$ as

$$A^2(t) = \tilde{A}^2 \sigma_X^2 + Y^2 \left(1 - \sigma_X^2 / \sigma_Y^2\right)$$

(6.81)

Some empirical evidence from a simulation study supports the assumption that the expected maximum resultant response $\mu_{A_n}$ occurs if $\tilde{A}(t)$ and $Y(t)$ simultaneously attain $g_G$ and $g_f \sigma_Y$.

Substituting $\tilde{A} = g_G$ and $Y = g_f \sigma_Y$ into Eq. (6.81), the expected maximum response of a combined process $A(t)$ can then be approximated as

$$\mu_{A_n} \approx \sqrt{\left(g_G^2 - g_f^2\right)\sigma_X^2 + g_f^2 \sigma_Y^2}$$

(6.82)

The use of Eq. (6.82) for predicting the peak resultant response can be named as the combined process (CP) method. Given the mean $E(A)$ and standard deviation $D(A)$ of the combined resultant response, the expected maximum resultant response can also be estimated using the Weibull peak factor as

$$\mu_{A_n} \approx E(A) + D(A) g_W$$

(6.83)

where $g_W$ is given in Eq. (6.66). For comparison purpose, using Eq. (6.83) with the Weibull peak factor $g_W$ for statistically determining the peak resultant acceleration is called as the statistical-based method. A conservative but simple method for estimating the peak resultant response is the direct combination of two component peaks as

$$\mu_{A_n} \approx \sqrt{g_f^2 \sigma_X^2 + g_f^2 \sigma_Y^2} = g_f \text{RMS}(A)$$

(6.84)

where $g_f$ denotes the Davenport's peak factor given in Eq. (6.59).

160

Using Eq. (6.84) for predicting the peak resultant response is regarded as the conventional method. It is worth noting that while the combined process method is only applicable to the resultant process, the statistical-based method is more general and applicable to any scalar random response process.

### 6.3.3 Time-variant reliability of a scalar response process

The Time-variant reliability of a scalar response process can be defined in terms of the probability distribution of the first-passage time and the corresponding maximum value distribution as discussed in section 6.2.2. Hence, the time-variant reliability can be readily calculated based on the specific information of a peak response distribution. In design practice, a design threshold criterion is given in accordance with the design code corresponding to a limit state for serviceability design or strength design. If a peak response of interest in buildings is treated as a random variable, a threshold value imposed on the random peak response may be exceeded by any peak response, chosen at random, with a very small probability. Therefore, the time-variant reliability of wind-induced dynamic response is equal to the probability of no barrier-crossing during a given time interval.

For a peak response following the Rayleigh distribution, the probability of success in the peak response not exceeding the expected maximum response determined by the Davenport's peak factor $g_f$ is given as

$$P_s(Y_m \leq g\sigma_Y) = F_{Y_m}(g_f\sigma_Y) = 1 - \exp\left[-\frac{1}{2}\left(\frac{g_f\sigma_Y}{\sigma_Y}\right)^2\right] = 1 - \exp\left(-\frac{g_f^2}{2}\right) \quad (6.85)$$

For instance, given a typical peak factor of $g_f = 3.50$, $P_s = 0.9978$ is obtained. Then the reliability index may be approximated by the first-order reliability method (FORM) as

$$\beta_R = \Phi^{-1}(P_s) = \Phi^{-1}\left[1 - \exp\left(-\frac{g_f^2}{2}\right)\right] \quad (6.86)$$

The reliability index is numerically evaluated by Eq. (6.86) and is shown in Figure 6.4 for $g_f = 3.0 \sim 4.5$. The linear regression relationship between the reliability index and the peak factor is given as

$$\beta_R \approx (g_f - 0.9223)/0.9056 \quad (6.87)$$

The above empirical equation may be used to estimate the reliability index given the value of

Davenport's peak factor.

If the expected maximum response is predicted by the statistical-based method, the reliability index corresponding to the use of the Weibull peak factor can be calculated as

$$\beta_W = \Phi^{-1}(P_s) = \Phi^{-1}\left[1 - \exp\left(-\frac{g_W^\kappa}{\rho}\right)\right] \qquad (6.88)$$

where $g_W$=the Weibull peak factor given in Eq. (6.66); $\kappa$ = the shape parameter of the Weibull peak distribution; $\rho$ = the scale parameter. The time duration considered is used for determining the sample size $n$ of peaks and troughs and in turn the value of peak factor.

If the expected largest value of a resultant response process is estimated by the combined process method, the reliability index can then be approximated as follows

$$\beta_G = \Phi^{-1}(P_s) = \Phi^{-1}\left\{1 - \frac{1+g_G^2}{2}\exp\left(-\frac{g_G^2+g_f^2-1}{2}\right)\right\} \qquad (6.89)$$

where $g_G$ = the Gamma peak factor given in Eq. (6.77); $g_f$= the Davenport's peak factor given in Eq. (6.59).

It is noted that Eq. (6.89) is calculated based on the assumption of independence between the component process and the reduced combined process. Similarly, the reliability index corresponding to the use of the conventional Davenport's peak factor for predicting a peak resultant acceleration can be calculated if assuming the independence of two component processes as

$$\beta_C = \Phi^{-1}(P_s) = \Phi^{-1}\left[1 - \exp\left(-g_f^2\right)\right] \qquad (6.90)$$

## 6.4    Application: The expected largest wind-induced response of tall buildings

### 6.4.1    Time history analysis for two tall buildings

Two tall building examples are used to illustrate the effectiveness and practical application of the statistical analysis technique on the prediction of largest wind-induced responses of tall buildings. One is the 60-story benchmark building as presented in Chapter 6, the other is the 45-story CAARC building as described in Chapter 4. The application of the probabilistic peak factor in Eq. (6.31), the Weibull peak factor in Eq. (6.66) and the Gamma peak factor in Eq. (6.77) is demonstrated. The first-order time-variant reliability of wind-induced response is also calculated using the approximation approach based on the first-passage probability of a scalar process.

Aerodynamic wind forces acting on two tall buildings were measured by the SMPSS technique in the wind tunnel using two 1:400 scale rigid models. A 10-year return period hourly mean wind speed of 34.7m/s at the reference height of 90 m was used to calculate wind-induced accelerations of the buildings. The 3-dimensional dynamic wind forces acting on the top level (heights ranged from 228 m to 240 m) of the 60-story benchmark building in the alongwind, crosswind and torsion directions are presented in Figures 6.5 (a, b, c), respectively. The time histories of wind forces given in Figures 6.5 (a, b, c) are corresponding to the 90-degree incident wind angle and the 10-year return period wind speed.

The modal superposition method provides a highly efficient and accurate procedure for performing time-history analysis of linear systems. A modal damping ratio of 1% is assumed for calculating the acceleration responses of the two buildings. The step-by-step Newmark integration method (Clough and Penzien, 1993; Chopra 2000) is employed in this study to perform the modal time history dynamic analysis of the buildings under wind excitation. As with any numerical-integration procedure the accuracy of this step-by-step method depends on the length of the time increment. In general, using a ratio of the time increment to the fundamental vibration period less than 10% gives reliable and stable numerical results (Chopra 2000). A time step of 0.25 second is used in the numerical computation of this study.

## 6.4.2   Results and discussion for peak factors

For simplicity, only the time history results of the 60-story benchmark building under 90-degree wind are presented in this section. The X- and Y-directional component acceleration and the corresponding resultant acceleration time histories under 90-degree wind perpendicular to the narrow face acting in the long direction of the building are presented in Figure 6.6 (a, b, c). The data histograms for the component acceleration and resultant acceleration processes are shown in Figure 6.7 (a, b, c). While Figure 6.7(a, b) demonstrate clearly a Gaussian distribution of the two component acceleration processes, Figure 6.7(c) indicates a non-Gaussian distribution for the resultant acceleration process. Based on the data samples of acceleration responses, the peak distribution and the largest peak distribution can be statistically estimated.

For example, considering Y-component acceleration response, 1301 data points of local maxima were chosen by using the POT approach from totally 14401 data points of the response history.

Figure 6.8(a, b, c) shows the fit of the Weibull distribution to peak acceleration responses of the building. By linear regression, a straight line, which is well fitted into the data points, is obtained. Such a good fitness indicates that the Weibull distribution offers a reasonable fitting to the peak acceleration data of the building, and is a good statistical estimation of the peak acceleration response distribution. Once the Weibull exponent $\kappa$ is available, the Weibull peak factor can be evaluated using Eq. (6.66).

The Weibull peak factors for component and resultant acceleration processes are plotted as a function of mean zero-crossing rate in Figure 6.9. The Weibull peak factors for the Y-component acceleration response are about 9% smaller than the Davenport's peak factor. Such a significant reduction can be explained by the shape of the Weibull peak distribution as shown in Figure 6.10. The Weibull peak distribution with $\kappa = 2.1763 > 2$ demonstrates a narrower and steeper tail than the Rayleigh distribution, which indicates that there are fewer largest peaks than that for the Rayleigh distribution. Therefore, the expected largest peak response arising from the peak response population following the Weibull distribution with $\kappa = 2.1763$ is smaller than that from the Rayleigh population. For the X-component and resultant acceleration processes, the shape parameters of the corresponding Weibull distribution are equal to 1.9264 and 1.8388, respectively. As shown in Figure 6.9, the Weibull peak factors with $\kappa < 2$ for the X-component and resultant accelerations are larger than the Davenport's peak factor. This means that they have more largest peaks than that for the Rayleigh distribution, as demonstrated by the Weibull curve with $\kappa = 1.9264$ and $\kappa = 1.8388$ in Figure 6.10. The analytical Gamma peak factor for the resultant acceleration response is also shown in Figure 6.9. It is found that the Gamma peak factor response gives the larger values than that of Davenport's peak factor and the corresponding Weibull peak factor for the resultant acceleration response.

In order to consider the spectral effects, it is necessary to evaluate the probabilistic peak factor using Eq. (6.31) with the probability of no exceedance $p=0.57$. The bandwidth parameter $q$ of the random acceleration response process is required by Eq. (6.31) to capture the spectral effects on the peak factor. Given the acceleration time histories, the power spectral density curves corresponding to the X-direction, Y-direction component and resultant acceleration processes can be obtained by the spectral analysis and are shown in Figure 6.11. For the PSD curve of Y-component acceleration, the first translational modal frequency (0.188 Hz) and the first two torsional modal frequencies (0.413 and 0.744 Hz) clearly indicates three resonant peaks, while the second translational mode (0.329 Hz with swaying along the long direction of the building) does not contribute noticeable resonant effects in the Y-direction acceleration response spectrum. A noticeable resonant peak corresponding to the second sway mode and other peaks to all other modes can be found in the spectrum of the X-component acceleration as shown in Figure 6.11. Also it is worth noting that the resonant peaks of the PSD curve of the resultant acceleration are shifted from the corresponding peak positions of the PSD curves of the component accelerations.

Based on the acceleration response spectra, various orders of spectral moments can be computed by numerical integration, and the bandwidth parameter defined by the spectral moments can be evaluated. The bandwidth measure $q$ is then found to be 0.2, 0.4 and 0.7 for X-directional, Y-directional component acceleration and resultant acceleration processes, respectively. The probabilistic peak factors with $q$=0.2, 0.4 and 0.7 are shown in Figure 6.1. It is found that the probabilistic peak factor with $q$=0.2 corresponding to the narrowest bandwidth gives the lowest values of probabilistic peak factor, and about 6% lower values than that of the Davenport's peak factor.

### 6.4.3 Results of the expected largest component acceleration response

For simplicity, two component accelerations of the 60-story benchmark building under 90-degree wind are considered in this section. Table 6.1 presents the largest component accelerations calculated by the conventional Davenport's peak factor method and the statistical-based Weibull peak factor method. The reliability indexes related to two different expected largest accelerations are also given in the table.

The peak factors given in the table are calculated based on a vibration duration of $\tau = 600s$. The mean-crossing rate of vibration for the 60-story building is approximated by its first modal frequency $v_0 = 0.186\text{Hz}$. Based on the two different peak factors, two methods can be used to estimate peak acceleration responses. For X-direction component, the conventional Davenport's peak factor method gives 3% lower value of peak acceleration than that obtained by the statistical-based Weibull peak factor method. The Y-directional peak acceleration, however, has been overestimated 8% by the conventional method. The peak acceleration response determined using the conventional method may be significantly overestimated or at times underestimated. Therefore it is necessary to carry out the statistical-based method whenever the time history response data is available. It is found that the reliability indexes related to the different values of peak accelerations are equal to the same value of 2.58 regardless of the methods used and the acceleration response components. The same reliability index results indicate that using the statistical-based method and the conventional method can attain the same level of reliability.

165

### 6.4.4 Results of the expected largest resultant acceleration response

In order to more reasonably predict the expected maximum acceleration response, it is recommended that the statistical-based method be used with the help of the Weibull peak factor if the time history response data of the building is available. For the resultant acceleration response, the combined process method using Eq. (6.82) provides a good alternative to predict the expected maximum acceleration if only the spectral analysis in the frequency domain is carried out without time history response data. Table 6.2 presents the wind-induced acceleration responses at the top corners of two buildings. The expected largest peak resultant acceleration is calculated using three different methods, i.e., the conventional method using Eq. (6.84) based on Davenport's peak factor $g_f$, the combined process method using Eq. (6.82) based on the Gamma peak factor $g_G$, and the statistical-based method using Eq. (6.83) with the aid of Weibull peak factor $g_W$. It is evidently shown in Table 6.2 that the conventional method has generally resulted in an overestimation of the expected maximum response of a combined resultant process, as much as 25% higher than the value predicted by the statistical-based method.

Consider the 60-story benchmark building under 90-dgree wind firstly. Given the RMS value of 13.6 milli-g for the resultant acceleration process, the expected peak resultant acceleration response is approximated as 44.3 milli-g by the conventional method. When using the statistical-based method, the expected maximum resultant acceleration response is found to be 39.9 milli-g, indicating a 10% reduction from 44.3 milli-g. The combined process method gives the peak resultant acceleration of 43.5 milli-g, which is only 2% lower than the value of 44.3 milli-g. Under 0-degree wind, the CP method and the statistical-based method also give respectively a 2% and 21% lower value than the resultant acceleration of 33.4 milli-g obtained by the conventional method. Since the Y-component acceleration has much larger standard deviation value than that of the X-component acceleration, the combined resultant process is actually dominated by the Y-component acceleration. Hence the combined probability distribution effect in the resultant process captured by the Gamma peak factor is not so significant that the CP method gives only 2% lower value than that obtained by the conventional method.

For the CAARC building under 0-degree wind, the statistical-based method gives a peak resultant acceleration of 17.7 milli-g, indicating a 25% reduction from the value of 23.5 milli-g predicted by the conventional method. When compared to the value of 19 milli-g obtained by the CP method, the conventional method also overestimates the peak resultant acceleration by nearly 20%. It is observed from the results that if two acceleration components have closer values of standard deviation, the peak resultant acceleration will be relatively more overestimated by the conventional method, in which the significant combined probability distribution effect in the resultant process for the CAARC building has not been captured.

The peak factors and the corresponding reliability indexes of the peak resultant acceleration responses of the two buildings are given in Table 6.3. As a conservative approach, the conventional method gives the highest level of reliability. For serviceability design, the reliability index higher than 4 may be too stringent and may result in the overly conservative design in practice (Ang and Tang 1975). Apparently, the expected maximum dynamic response may be predicted by the combined process method or the statistical-based method with a more reasonable level of reliability for serviceability design, i.e., reliability index being within from 2 to 3. While the combined process method gives a reliability level slightly below 4, the statistical-based method attains a reliability index of around 2.6. It is believed that various methods with different reliability implication on predicting the expected largest peak response can be useful in the reliability performance-based serviceability design, in which the different reliability levels corresponding to various serviceability design requirements can be specified.

## 6.5 Summary

Time-variant reliability analysis method of wind excited buildings has been conducted in the context of random vibration. The important properties of random response process, such as the mean level-crossing rate and peak distribution, have been revisited by using the statistical approach. Various peak factor approaches have been studied in detail. The influences of the mean level-crossing rate and peak distribution on the estimation of peak factors have been extensively investigated for the component acceleration process and the resultant acceleration process. The numerical results indicate that the use of Davenport peak factor generally yields a conservative estimate of the expected largest response for a narrowband response process. A new more refined formula for the probabilistic peak factor is developed and explicitly expressed as a function of excitation duration, the probability of exceedance and the shape parameter (bandwidth) of the power spectral density of the underlying random response process. The probabilistic peak factor can be directly related to an explicit measure of reliability (or the probability of the largest peak responses without exceeding specified certain design threshold value) for structural design.

Based on the asymptotic theory of statistical extremes, the expected Weibull peak factor can be derived from the generalized Weibull peak distribution. The expected Weibull peak factor has a shape parameter, which can be determined by statistical analysis using the peak response data sample. For a combined Rayleigh resultant process, a closed-form peak factor, the so called Gamma peak factor, has been devised for estimating the expected largest peak resultant accelerations of wind sensitive tall buildings without the need of time-history response analysis. The relationship of various peak factors to the first-order reliability index has been explicitly expressed given the underlying local maximum response distribution.

The 60-story benchmark building and the 45-story CAARC building have been utilized to illustrate the use of various peak factors and the corresponding methods for predicting the expected maximum acceleration response and determining the reliability index for the predicted response. The results indicate that the expected maximum acceleration response determined using the conventional method may be significantly overestimated or at times underestimated. When predicting the expected maximum component acceleration response, the statistical-based method can attain the same level of reliability as the conventional method. The combined process method using the Gamma peak factor can provide a more reasonable estimation of the expected maximum resultant acceleration than the conventional method, especially when two component accelerations have the close values of standard deviation. Unlike the statistical-based method, the combined process method as well as the conventional method do not require time history response analysis, which is computationally more expensive than the frequency domain spectral analysis. Since the different methods have various reliability implications, the choice of method may depend on the reliability requirements as needed in a performance-based serviceability design.

**Table 6.1  Component accelerations at the top corner of the 60-story building**

| Component | Standard deviation | Peak factors | | Peak acceleration | | Reliability index | |
|-----------|--------------------|--------------|--------------|----------------------|----------------------|-------------------|-----------------|
| | | $g_f$ | $g_W$ | Conventional method | Statistical method | $\beta_R$ | $\beta_W$ |
| X-direction | 2.7 | 3.2588 | 3.3878 | 8.8 | 9.1 | 2.58 | 2.58 |
| Y-direction | 13.3 | 3.2588 | 3.0019 | 43.3 | 39.9 | 2.58 | 2.58 |

**Table 6.2** Acceleration responses at the top corner of the 60-story building and the 45-story CAARC building

| Wind direction | Component (milli-g) | | Resultant (milli-g) | | Peak resultant (milli-g) | | |
|---|---|---|---|---|---|---|---|
| | $\sigma_X$ | $\sigma_Y$ | Mean | RMS | Conventional method | Statistical method | CP Method |
| 0-degree wind | 2.4 | 10 | 8.6 | 10.3 | 33.4 | 26.4 (-21%) | 32.8 (-2%) |
| 90-degree wind | 2.7 | 13.3 | 10.9 | 13.6 | 44.3 | 39.6 (-11%) | 43.5 (-2%) |
| 0-degree wind (CAARC) | 5.1 | 5.1 | 6.3 | 7.2 | 23.5 | 17.7 (-25%) | 19.0 (-19%) |

Note: The numbers in brackets represent the percentage reduction of resultant acceleration values calculated by CP method or Statistical method as compared with the value obtained by conventional method.


**Table 6.3** Peak factors and reliability index for resultant acceleration responses of the two buildings

| Wind direction | Peak factors | | | Reliability index | | |
|---|---|---|---|---|---|---|
| | $g_f$ | $g_W$ | $g_G$ | $\beta_C$ | $\beta_W$ | $\beta_G$ |
| 0-degree wind | 3.2588 | 3.1722 | 3.7325 | 4.06 | 2.58 | 3.86 |
| 90-degree wind | 3.2588 | 3.5624 | 3.7325 | 4.06 | 2.58 | 3.86 |
| 0-degree wind (CAARC) | 3.2763 | 3.3859 | 3.7508 | 4.09 | 2.60 | 2.29 |

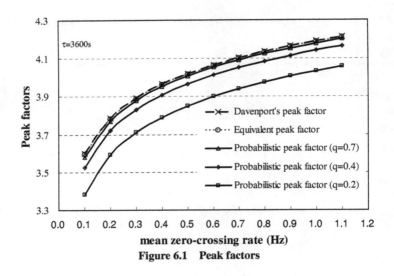

**Figure 6.1    Peak factors**

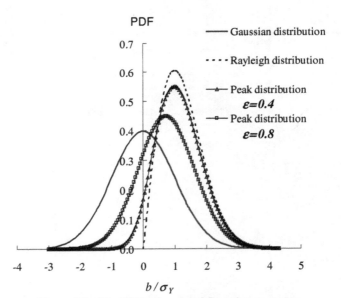

**Figure 6.2    The PDFs of peaks of Gaussian processes**

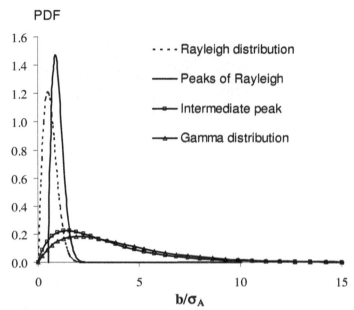

**Figure 6.3    The PDFs of peaks and the intermediate peak variable**

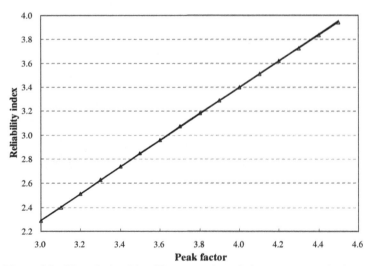

**Figure 6.4    The relationship of Davenport's peak factor to reliability index**

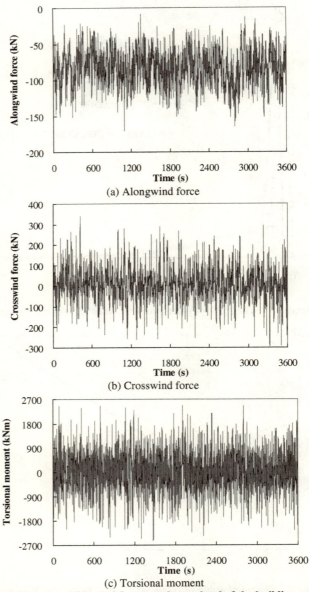

(a) Alongwind force

(b) Crosswind force

(c) Torsional moment

**Figure 6.5    Time histories of 3D wind forces at the top level of the building under 90-degree wind**

172

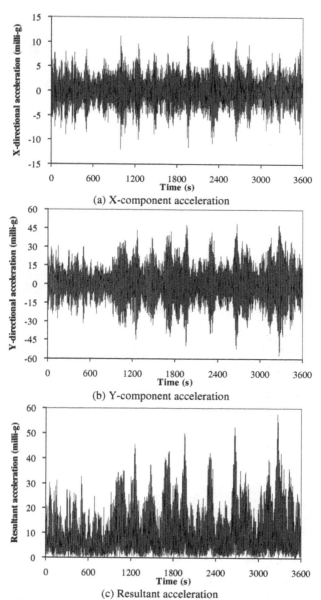

(a) X-component acceleration

(b) Y-component acceleration

(c) Resultant acceleration

**Figure 6.6    Acceleration time histories at the top corner of the building under 90-dgree wind**

173

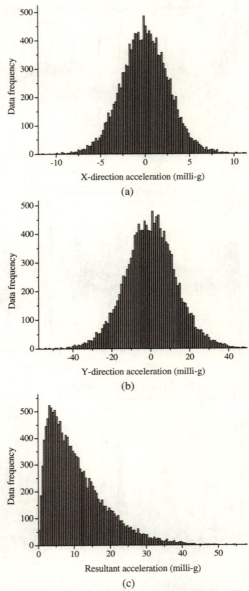

**Figure 6.7    Acceleration histograms of the 60-story building**

174

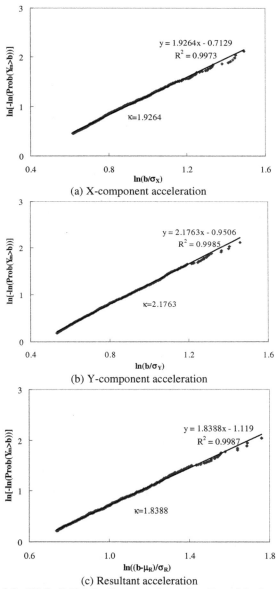

(a) X-component acceleration

(b) Y-component acceleration

(c) Resultant acceleration

**Figure 6.8    Fit of the Weibull distribution to peak acceleration of the benchmark building**

175

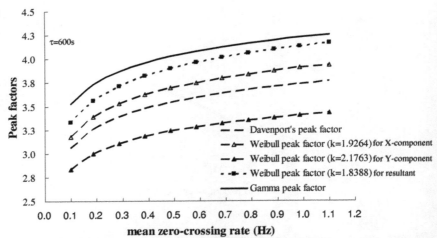

**Figure 6.9** The Weibull and Gamma peak factors for component and resultant accelerations of the 60-story building

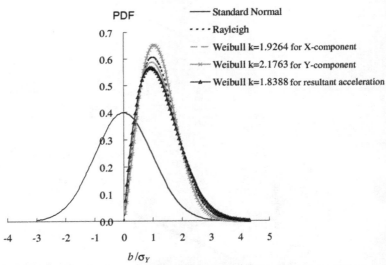

**Figure 6.10** The Weibull distributions of component and resultant accelerations of the 60-story building

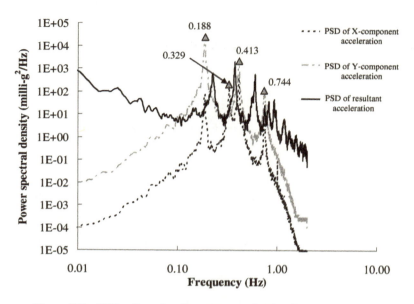

**Figure 6.11 PSDs of acceleration responses for the 60-story building**

177

# CHAPTER 7    Mathematical Formulation of Reliability-based Performance Design Optimization

## 7.1    Formulation of probabilistic design optimization problem

For a general structure with random design variables and subjected to stochastic excitation, the optimization problem can be stated as:

Find  $\mathbf{D} = \mathbf{D}^*$ , which minimizes  $W(\mathbf{D})$

where D denotes the mean values of the corresponding random design variables or deterministic design variables of structural elements.

subject to

$$P\left\{\bigcup_{h=1}^{l}[g_{hj}(\mathbf{Y}(\mathbf{D},\mathbf{u},t)) \geq B_{hj}]\right\} \leq P_j, \quad 0 \leq t \leq T, \mathbf{u} \in U \subset R^n, j = 1, 2, ..., m_1 \quad (7.1)$$

$$E[g_j(\mathbf{Y}(\mathbf{D},\mathbf{u},t))] \leq \alpha_j, \quad 0 \leq t \leq T, \mathbf{u} \in U \subset R^n, j = m_1 + 1, ..., m_2 \quad (7.2)$$

$$g_j\{\mathbf{Y}(\mathbf{D},\mathbf{u},t)\} \leq \alpha_j, \quad 0 \leq t \leq T, \mathbf{u} \in U \subset R^n, j = m_2 + 1, ..., m_3 \quad (7.3)$$

$$P\left\{g_j(\mathbf{D}) \geq B_j\right\} \leq P_j, \quad j = m_3 + 1, ..., m_4 \quad (7.4)$$

$$g_j(\mathbf{D}) \leq \alpha_j, \quad j = m_4 + 1, ..., m_5 \quad (7.5)$$

where $t$ denotes time variable; $u$ denotes spatial coordinates; $g_{hj}$ denotes an evaluation function of the random response vector $\mathbf{Y}(\mathbf{D},\mathbf{u},t)$ of the structure in terms of spatial variation index $h$; $g_j$ denotes a utility function of $\mathbf{Y}(\mathbf{D},\mathbf{u},t)$ or random design variables $\mathbf{D}$; $B_{hj}$ and $B_j$ denote the limits on the performance function $g_{hj}$ and $g_j$ respectively; $P_j$ is the specified failure probability in the $j$-th failure mode; $a_j$ denotes the assigned limiting value for any utility function of design variables.

The constraints from (7.1) to (7.3) were written down to take into account the specific design requirements on the deterministic or probabilistic characteristics of the random responses of structure systems. These constraints must be satisfied for systems at any time and any location. The constraints (7.4) and (7.5) impose the design limits on the deterministic or probabilistic characteristics of systems and their components.

The above constraints can be separated into three categories: one representing the time and spatial depending constraints arising from random vibration, including expressions (7.1) to (7.3); especially, constraint expression in (7.1) can be called as reliability constraint; (7.2) is response moment or cumulument constraint; (7.3) denotes dynamic response constraint. The next category of constraints in (7.4) is representing the probabilistic constraints involving only the random design variables. The third category in (7.5) is representing the deterministic and static constraints.

## 7.2 Treatments of constraints

### 7.2.1 Probabilistic constraints involving time and spatial parameter

The primary difficulty of the optimization problem is the treatment of numerous and various constraints through the whole design process, especially the time and spatial depending constraints. The probabilistic constraints involving time and spatial parameter arising due to the random vibration environment have been stated in the equations of (7.1) to (7.3). For serviceability design of tall buildings, the function $g_{hj}$ in Eq. (7.1) may represent the acceleration or drift at elevation $h$ of a building structure with the subscript $j$ referring to the failure mode corresponding to the acceleration response of the building structure under the certain level of extreme wind or seismic excitation. From the ultimate strength viewpoint, the quantity of $g_{hj}$ may represent the stress in the $h$-th component of a structure due to $j$-th case of loading. It is worth to point out that it is often the serviceability limit state which governs the design of high-rise structures under wind loadings.

Denoting the failure event due to outcrossing as

$$E_{hj} = \{ g_{hj}(\mathbf{Y}(\mathbf{D},\mathbf{u},t)) \ge B_{hj} \}$$
$$0 \le t \le T, \mathbf{u} \in U \subset R^n, j = 1,2,...,m_1 \tag{7.6}$$

Then the constraint (7.1) can be rewritten as

$$P\left\{ \bigcup_{h=1}^{l} E_{hj} \right\} \le P_j, \quad j = 1,2,...,m_1 \tag{7.7}$$

The above constraints may be regarded as the reliability-based constraints, which describe the system failure probability due to a series of component failure events in different failure modes. Actually in terms of system reliability, structural systems may be divided into two classifications,

179

i.e. series and parallel systems. A series system fails to perform when any of its components or modes fail and survives only all components or modes survive. A parallel system fails only if all components and modes fail. The system reliability considered later in this monograph belongs to series system. Hence the failure of a building system is defined by the first failure of a component or a loss of serviceability at any location.

Theoretically, in order to calculate the joint probability expressed in (7.7), the joint probability density function of multiple failure events should be known and high dimensional integration have to be evaluated. In order to reduce the computation efforts, some approximation approaches to quantify the joint probability should be pursued. It can be assumed that the most critical position $u^*$ can be determined for each failure event $E_{hj}$. It is known that the cumulative distribution function of failure probability is not a descent function. The maximum probability of failure would occur at the first-passage time $t = T$. Hence the spatial and time parameters can be eliminated as follows

$$P\{E_{hj}\} = F_{hj}(\mathbf{D}, u^*, T) = F_{hj}(\mathbf{D}, T) \tag{7.8}$$

where $F_{hj}(\mathbf{D},T)$ denotes the failure probability for the $h$-th component or position in the $j$-th failure mode. The bounds of the joint failure probability can be established based on the dependence of individual failure events. It is known that if all $E_{hj}$ are completely correlated to each other, one obtain

$$P\left\{\bigcup_{h=1}^{l} E_{hj}\right\} = \max_{h}\{P(E_{hj})\} = \max_{h}\{F_{hj}(\mathbf{D}, T)\} \tag{7.9}$$

If all failure events of $E_{hj}$ are completely uncorrelated

$$P\left\{\bigcup_{h=1}^{l} E_{hj}\right\} = \sum_{h=1}^{l}\{P(E_{hj})\} = \sum_{h=1}^{l}\{F_{hj}(\mathbf{D}, T)\} \tag{7.10}$$

Since, in reality, the events $E_{hj}$ are partially correlated, the failure probability of the union of events is bounded by the probabilities of the extreme cases of complete correlation and no correlation of events $E_{hj}$. Hence, the following inequality can be obtained as

$$\max_{h}\{F_{hj}(\mathbf{D}, T)\} \le P\left\{\bigcup_{h=1}^{l} E_{hj}\right\} \le \sum_{h=1}^{l}\{F_{hj}(\mathbf{D}, T)\} \tag{7.11}$$

By using the above inequality, the left-side of expression (7.1) may be replaced by $\sum_{h=1}^{l}\{F_{hj}(\mathbf{D}, T)\}$

at the conservative side, such that the time and spatial dependence of the constraint (7.1) was eliminated.

Similarly, the constraints (7.2) and (7.3) can be treated to eliminate the time and spatial dependence. Using the operation of mathematic supremum in time and space domain, the left sides of constraints (7.2) and (7.3) can be replaced respectively as following

$$\sup_{(t,\mathbf{u})}\{E[g_j(\mathbf{Y}(\mathbf{D},\mathbf{u},t))]\} \le \alpha_j, \; j = m_1 + 1,...,m_2 \tag{7.12}$$

$$\sup_{(t,\mathbf{u})}\{g_j(\mathbf{Y}(\mathbf{D},\mathbf{u},t))\} \le \alpha_j, \; j = m_2 + 1,...,m_3 \tag{7.13}$$

To sum up, probabilistic constraints involving time and spatial parameters, expressed in (7.1) to (7.3), were greatly simplified by eliminating the time and spatial parameters in the constraints. For sloving the reliability-based optimization problem, $F_{hj}$ ($\mathbf{D}$, $T$), denoted as the failure probability, need to be further determined.

In the circumstance of random vibration, the specific function $F_{hj}$ ($\mathbf{D}$, $T$) to measure the probability of failure related to a given failure mode can be classified into three categories:

(i) cumulative fatigue damage;

(ii) first excursion failure;

(iii) failure due to excessive vibration.

By introducing some assumptions, several analytical results on the estimation of failure probability corresponding to above three failure modes have been available in the literatures (Lin 1967).

### 7.2.2 Probabilistic constraints involving random design variables

Considering the constraint (7.4) for the reliability-based design of structure subjected to static loads. For instance, here $g_k$ may represent the fundamental natural frequency of the structural system. Actually, in the dynamic performance design of engineering structures, the frequency constraints are very important. By limtting the natural frequencies of the structure operating in a random vibration environment to lie in regions where the power spectral density of excitation has small values, the dynamic response of the structure can be significantly reduced thereby implying a better design.

Since the design variables are random, the material properties and in turn the dynamic characteristics of structure system are also random. Considering the general cases in terms of load effect ($S$) and resistance ($R$), the constraint (7.4) can be rewritten as

$$P[S_j(\mathbf{D}) > R_j(\mathbf{D})]\} \leq P_j, j = m_3 + 1,..., m_4 \tag{7.14}$$

which means that the probability of the $j$-th load effect, $S_j$, exceeding the corresponding resistance, $R_j$, must be less than or equal to a given acceptance failure probability, $P_j$. If the random design variables are jointly normally distributed, the value of $P_j$ may be related to the new random variable so call safety margin, defined by $Z_j = R_j - S_j$, as following

$$P_j = P(Z_j < 0) = 1 - \Phi(\beta_j), \quad j = m_3 + 1,..., m_4 \tag{7.15}$$

where $\Phi(\beta_j)$ is the standard normal distribution corresponding to the reliability index $\beta_j$ as

$$\beta_j = \frac{\mu_{R_j} - \mu_{S_j}}{\sqrt{\sigma_{R_j}^2 + \sigma_{S_j}^2}} \tag{7.16}$$

where $\mu_R$ and $\mu_S$ are mean values, and $\sigma_R^2$ and $\sigma_S^2$ are variances of random variables $R_j$ and $S_j$ respectively. Traditionally, in terms of structure design, the condition $\mu_R > \mu_S$ should always be satisfied. Based on reliability index approach (Frangopol and Maute 2003), and considering $\mu_R > \mu_S$, the constraint (7.14) can be replaced by

$$\mu_S + \beta_j^L \sqrt{\sigma_R^2 + \sigma_S^2} - \mu_R \geq 0 \tag{7.17}$$

where $\beta_j^L$ denotes the lower bound of reliability index corresponding to the given failure probability $P_j$. The one to one relationship between $\beta_j^L$ and $P_j$ was given by Eq. (7.15), and was schematically shown in Figure 7.1. Depending on the specific problem of interest, the proper evaluations of mean values and variances of random variables $R_j$ and $S_j$ need to be further considered.

### 7.2.3 Deterministic constraints

The constraints (7.5) are the deterministic form of constraints, which have been widely used in the conventional optimization approach. $g_j(\mathbf{D})$ is a deterministic response quantity and $\alpha_j$ is the specified upper bound on $g_j(\mathbf{D})$.

## 7.3 Basic solution

After employing the treatments of constraints as presented in the previous section, the reliability-based optimization problem described in the section 7.1 can be reduced to a nonlinear programming problem in terms of explicitly constraint formulations with random design variables as following:

$$\text{Find } \mathbf{D} = \mathbf{D}^*, \text{ which minimizes } W(\mathbf{D}) \qquad (7.18)$$

where $\mathbf{D}$ denotes the mean values of the random design variables.

subject to

$$g_j(\mathbf{D}) \le \alpha_j, \quad j = 1, 2, ..., n \qquad (7.19)$$

where $g_j(\mathbf{D})$ represents any constraints in terms of functions, including specific probability density functions; $\alpha_j$ denotes any specified limit or threshold value for the corresponding constraint. In this form it is possible to solve the design optimization problem by mathematical programming techniques (Fiacco et al. 1990) or Optimality Criteria method (Chan 2001, Chan 2004), which has been presented in Chapter 4.

Some available mathematical programming techniques also can deal with such nonlinear programming problems. The penalty function method has been applied to solve many nonlinear optimization problems. In this method, nonlinear programming problems would be converted into a sequence of unconstrained minimization problems. The general form of penalty functions is expressed as

$$\phi(\mathbf{D}, r_k) = W(\mathbf{D}) + r_k \sum_{j=1}^{M} G_j\left(g_j(\mathbf{D})\right) \qquad (7.20)$$

where $G_j$ is a function of $g_j$; $r_k$ is a decreasing sequence so that $r_{k+1} < r_k$. For any fixed value of $r_k$, the minimization is performed by taking successive steps in a number of different directions as

$$\mathbf{D}_{i+1} = \mathbf{D}_i + \tau_i \mathbf{S}_i \qquad (7.21)$$

where $\mathbf{D}_{i+1}$ is the design vector corresponding to the minimum of the penalty function along the search direction $\mathbf{S}_i$; $\tau_i$ is the minimizing step length and $\mathbf{D}_i$ is the starting design vector for the $i$-th iteration. It has been proved that the solution of $\phi(\mathbf{D}, r_k)$ when $k$ approaching infinite would converge into the solution of initial nonlinear programming problem (Fiacco et al. 1990). So that the penalty function method is also known as sequential unconstrained minimization technique.

## 7.4 Wind-induced performance based design optimization

For simplification purpose, our problem about wind-induced performance based design optimization will focus more on wind loading uncertainty under random vibration environment. Hence the distributions of design variables, like material properties and dynamic properties of building systems, will be assumed to be normal or lognormal. The design variables used in this monograph can be regarded as mean values if without any specific claim. The stochastic nature of wind exciting process and the uncertainty of extreme wind events or wind hazards would be incorporated into the synthesis of design optimization.

### 7.4.1 Reliability-based design optimization

The formulation of the design optimization problem can be exactly expressed as from (7.1) to (7.5). In terms of wind-resistant design, wind-induced response would be constrained during the design optimization process. The constraint of wind-induced drift would be formulated as a deterministic constraint here. On the other side, the constraint related to wind-induced vibration should be treated as the probability constraints.

The constraint of wind induced acceleration can be treated as first excursion failure probability constraint as mentioned in previous section. The first excursion probability of the random acceleration response process $g_{hj}(\mathbf{Y}(\mathbf{D}, u^*, t)) = \ddot{\mathbf{Y}}_{hj}(\mathbf{D}, u^*, t)$ in the time interval $0 \leq t \leq T$ can be defined as ($u^*$ representing the position of floor corner)

$$F_{hj}(T, a_j^U) = P(\sup_{0 < t \leq T} \ddot{\mathbf{Y}}_{hj}(t)) \geq a_j^U) \tag{7.22}$$

where $h$ denotes the story number of the tall building; $j$ denotes the intensity level of extreme wind event under consideration; $a_j^U$ denotes the acceptance acceleration criteria corresponding to $j$-th intensity level of wind. The above failure probability may be considered to be equivalent to

$$F_{hj}(T, a_j^U) = 1 - P(N(T) = 0 \mid \ddot{\mathbf{Y}}_{hj}(0) < a_j^U) P(\ddot{\mathbf{Y}}_{hj}(0) < a_j^U) \tag{7.23}$$

where $\ddot{\mathbf{Y}}_{hj}(0) < a_j^U$ signifies that the process $\ddot{\mathbf{Y}}_{hj}(t)$ starts in the safe domain at zero time; $N(t)$ is the number of excursions during the time interval $0 \leq t \leq T$.

The general solution of Eq. (7.23) is rather difficult to obtain owing to the need to account for the complete information of the stochastic process $\ddot{\mathbf{Y}}_{hj}(t)$. Fortunately, for reliability problems, excursions usually occur so rarely that it is often satisfactory for the individual excursions to be assumed independent events. The probability of no excursions in [0, T] may then be approximated using the Poisson distribution with zero events:

$$P(N(T) = 0) = \frac{(v^+ T)^0}{0!} \exp(-v^+ T) = \exp(-v^+ T) \tag{7.24}$$

where $v^+$ is the expected rate of up crossings of the given level $\ddot{\mathbf{Y}}_{hj}(t) = a_j^U$. It is also reasonable to assume:

$$P(\ddot{\mathbf{Y}}_{hj}(0) < a_j^U) = 1 \tag{7.25}$$

Substituting Eq. (7.24) and (7.25) into (7.23), one obtain

$$F_{hj}(T, a_j^U) = 1 - \exp(-v^+ T) \tag{7.26}$$

Assuming $\ddot{\mathbf{Y}}_{hj}(t)$ is a stationary Gaussian process with zero mean, the expected rate of up crossings can be obtained as (Rice 1944)

185

$$v^+ = \frac{\omega}{2\pi} \exp(-\frac{\left(a_j^U\right)^2}{2\sigma_{\ddot{Y}}^2}) \tag{7.27}$$

where $\omega$ is the natural circular frequency of the building system; $\sigma_{\ddot{Y}}$ denotes the standard deviation of random process $\ddot{Y}_{hj}(t)$. Using Eq. (7.27) in Eq. (7.26), the probability constraints for wind-induced acceleration can be written as

$$F_{hj}(T, a_j^U) = 1 - \exp\left[ -\frac{T\omega}{2\pi} \exp(-\frac{\left(a_j^U\right)^2}{2\sigma_{\ddot{Y}}^2}) \right] \leq P_j \tag{7.28}$$

If considering two-sided barrier to control negative acceleration, the first excursion probability can be defined as

$$F_{hj}(T, -a_j^U, a_j^U) = P(\sup_{0<t\leq T} \ddot{Y}_{hj}(t)) \geq a_j^U) \bigcup P(\inf_{0<t\leq T} \ddot{Y}_{hj}(t)) \leq -a_j^U) \tag{7.29}$$

where **inf** represents the mathematic infimum.

Considering the expected rate of down crossings v- is equal to the expected rate of up crossings v+, a good approximation for Eq. (7.29) is

$$F_{hj}(T, -a_j^U, a_j^U) = 1 - \exp\left[ -\frac{T\omega}{\pi} \exp(-\frac{\left(a_j^U\right)^2}{2\sigma_{\ddot{Y}}^2}) \right] \leq P_j \tag{7.30}$$

Actually, from the Eqs. (7.26) and (7.27), the cumulative distribution of the peak absolute response, defined as

$$\eta = \max_T |Y| \tag{7.31}$$

can be expressed as

$$F(\eta) = \exp(-\nu_0 T \exp(-\frac{1}{2}\tilde{\eta}^2)) \tag{7.32}$$

where $\tilde{\eta} = (\pi\eta - \mu_Y)/\sigma_Y; \nu_0 = \frac{1}{\pi}\frac{\sigma_{\dot{Y}}}{\sigma_Y}$, and can be approximated as the modal frequency of the structure system.

186

The peak factor widely used in wind engineering was derived from the cumulative distribution of the peak response, as the following by Davenport (1964):

$$g = \sqrt{2\ln f_0 \cdot T} + \frac{0.577}{\sqrt{2\ln f_0 \cdot T}} \qquad (7.33)$$

where $f_o$ is the modal frequency of the building. Let $\eta = \mu_Y + g\sigma_Y$ in Eq. (7.32), the probability of exceeding mean peak response can be calculated by

$$P(\eta > (\mu_Y + g\sigma_Y)) = 1 - \exp(-\nu_0 T \exp(-\frac{1}{2}g^2)) \qquad (7.34)$$

For instance, considering a typical tall building with a height of 120m, the fundamental frequency can be estimated as 46/120=0.383 Hz. Using 0.383 Hz as $\nu_0$ in Eqs. (7.33) and (7.34), one obtains the conditional probability of exceeding mean peak response is about 42%.

It is noted that the estimation of conditional failure probability using Eq. (7.28) or (7.30) is based on the assumption of independent barrier crossing events. Vanmarcke (1975) proposed a modified Poisson model taking into account the dependence between two successive crossing events to calculate the probability of no crossing during the interval (0,$T$). Using his formulation, the cumulative distribution of the peak absolute response can be written as

$$F(\eta) = \left[1 - \exp(-\frac{1}{2}\tilde{\eta}^2)\right]\exp\left[-\nu_0 T\frac{1 - \exp(-\delta^{1.2}\tilde{\eta}\sqrt{\pi/2})}{\exp(\tilde{\eta}^2/2) - 1}\right] \qquad (7.35)$$

where $\delta$ denotes spectral shape factor, which can be measured by spectral moments (Vanmarcke 1975). Based on the above distribution, Kiureghian (1980) proposed a revised formulation for peak factor as

$$g = \sqrt{2\ln(\nu_e \cdot T)} + \frac{0.577}{\sqrt{2\ln(\nu_e \cdot T)}} \qquad (7.36)$$

where $\nu_e = 1.63(\delta^{0.45} - 0.38)\nu_0$, $\delta < 0.69$; $\nu_e = \nu_0$, $\delta \geq 0.69$.

## 7.4.2 Performance-based design optimization

Recall the expression for annual probability of exceedance of a structural response parameter ($Y$)

exceeding a certain level ($b$)

$$P(Y > b) = \sum_{\text{all } d_k} P(Y > b \,|\, \text{given hazard intensity}=d_k)P(d_k) \qquad (7.37)$$

Where $P(d_k)$ denotes the annual probability of occurrence of the given extreme wind event with hazard intensity $d_k$.

For assessment of wind-induced performance, hazard intensity $d_k$ may be related to the extreme wind velocity of interested. $P(d_k)$ could be related to the return periods of extreme wind hazard. Actually, from the knowledge of extreme value distribution, annual extreme wind speeds can be described by Gumbel distribution in terms of cumulative distribution function as

$$P(V_{ext} \le v) = \exp\{-\exp[-\alpha(v-U)]\} \qquad (7.38)$$

Where $v$ denotes specific wind speed under consideration; $V_{ext}$ denotes annual largest wind speeds; $U$ is the mode; $a$ is a measure of the dispersion of the distribution. $1/a$ is known as a scale wind speed. $U$ and $a$ can be evaluated by regression analysis using available meteorological data.

Considering the extreme wind with a return period $R_k$, the annual probability of occurrence of the such extreme wind related to the wind hazard with hazard intensity $d_k$ can be expressed as

$$P(d_k) = 1/R_k \qquad (7.39)$$

From Eqs. (7.38) and (7.39), one obtain

$$P(d_k) = P(V_{ext} > v) = 1 - \exp\{-\exp[-\alpha(v-U)]\} = 1/R_k \qquad (7.40)$$

Given specific wind speed under consideration $v$ corresponding to $d_k$, the probability of $P(Y > b \,|\, \text{given hazard intensity}=d_k)$ also can be evaluated by assuming some statistical distribution of building performance involved.

For instance, if $Y$ representing acceleration response and $b$ representing acceptance criteria of acceleration, the probability of $P(Y > b \,|\, \text{given hazard intensity}=d_k)$ can be calculated as a first excursion problem

$$P(Y > b \mid \text{given hazard intensity}{=}d_k) = 1 - \exp\left[-\frac{T\omega}{2\pi}\exp(-\frac{b^2}{2\sigma_Y^2})\right] \quad (7.41)$$

Finally, the probability of Eq. (7.37) can be calculated based on the results of Eq. (7.40) and Eq. (7.41) considering different levels of wind velocity corresponding to different return period.

Replacing $d_k$ by $v$, the continuous form of Eq. (7.37) can be written as

$$P(Y > b) = \int_v P(Y > b \mid \text{given hazard intensity}{=}v)dP(v) \quad (7.42)$$

where $P(v) = 1 - \exp\{-\exp[-\alpha(v-U)]\}$.

The above approach is based on total probability theorem, which recognizes that there is some potential that extreme wind having a lower return period could result in the failure events of $Y > b$ and similarly, there is some potential that extreme wind having a higher return period could result in the lower probability of the failure events.

Till now, the constraints and their assessments involved in the framework of performance based design have been established. Next task is how to integrate the probabilistic constraints into optimization procedure to solve the optimum design problem in practice.

For simplification, it is useful to separate the design variables into deterministic variables and random variables. Let's denote $A_i$ as deterministic design variables, for instance, sizes of structure elements; Denoting $R_i$ as random design variables, for instance, extreme wind speed or structural damping involved in the procedure of dynamical analysis.

Accordingly, the wind-induced performance based design problem can be decomposed into two nested optimization problems based on two sets of design variables. One is the traditional deterministic optimization problem to consider the minimization of a predefined cost function; the other is the probabilistic optimization problem to take into account the uncertainties involved in the random vibration. These two optimization problems can be explicitly formulated as following respectively:

(1) Deterministic sizing optimization for minimizing cost function

Find $A_i = A_i^*$, which minimizes $Cost(A_i)$

subject to

static drift constraints: $\quad d_j \le d_j^U, \quad j = 1, 2, ..., N_j$

peak acceleration constraints: $\quad a_j \le a_j^U, \quad j = N_j + 1, 2, ..., N_g$

element sizing constraints: $\quad A_i^L \le A_i \le A_i^U, \quad i = 1, 2, ..., N$

(2) Probabilistic design optimization for minimizing wind-induced response

Find $\mathbf{R} = \mathbf{R}^*$, which minimizes acceleration $a_k(A_{i0}, \mathbf{R})$

where $\mathbf{R} = (r_1, ..., r_n)^T$; $k$ denotes the $k$-th case of acceleration under consideration; $A_{i0}$ indicates the initial value of deterministic design variable.

subject to

reliability constraints: $\quad \beta_{\text{target},k} = \sqrt{\tilde{\mathbf{R}}^T \tilde{\mathbf{R}}}$

where $\quad \tilde{\mathbf{R}} = (\tilde{r}_1, ..., \tilde{r}_n)^T$, $\tilde{r}_i = \dfrac{r_i - \mu_{r_i}}{\sigma_{r_i}}$, $i = 1, 2, ..., n$.

The relationship between these two sub problems, demonstrated in the flow chart Figure 7.2, is that the optimized distribution parameters (i.e., mean value) of random design variables of the second problem would be provided as the input for the first deterministic optimization problem. The detail implementation of reliability-based performance design optimization framework for tall buildings would be presented in the next chapter by defining a occupant comfort performance.

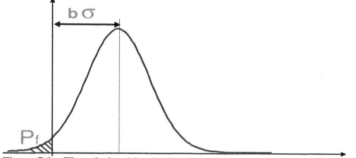

**Figure 7.1    The relationship of reliability index to the failure probability**

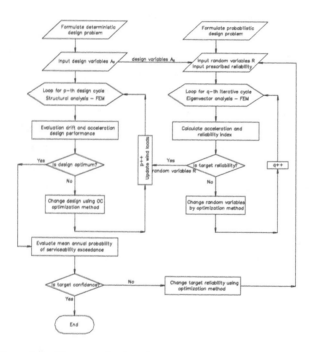

**Figure 7.2    Framework of performance based design optimization**

# CHAPTER 8   Reliability Performance-based Design Optimization of Wind-excited Tall Buildings

## 8.1   Introduction

In Chapters 6, the performance of tall buildings subjected to random wind excitations has been investigated in the time domain using statistical methods. The statistical analysis for peak response distribution provides an effective tool to accurately assess the time-variant reliability of tall buildings against wind-induced motions. The integrated time-variant reliability analysis and mathematical formulation of reliability-based performance design optimization framework is presented in Chapter 7. Although the time domain approach is capable of capturing the inherent uncertainty in random vibration, the other major uncertainties of building systems have not been formally treated in Chapter 6. This chapter aims to investigate the major uncertainties involved in building systems and design wind speeds and their effects on the prediction of wind-induced dynamic acceleration response. Two kinds of uncertainties may be associated with physical building systems and their external loading environments that are inherently random or with predictions and estimations of reality performed under conditions of incomplete or inadequate information. One kind is known as aleatory uncertainty, which arises from the fluctuating pressure in an anisotropic aerodynamic boundary turbulence field. Another kind of uncertainty due to lack of sample data or modeling knowledge is called as epistemic uncertainty. It is obvious that how all these uncertainties are modeled and represented in a tall building design has a direct impact on the reliability assessment of the design or the reliability-based design optimization (RBDO).

Although it is difficult to reduce the aleatory uncertainty directly, the time-variant reliability of a tall building against wind-induced random vibration can be improved by stiffness optimization. Unlike the aleatory uncertainty, the epistemic uncertainties associated with prediction or modeling error may be reduced through the use of more accurate models or the acquisition of additional sampling data. In this chapter, the epistemic uncertainties are formally treated in the reliability performance-based design optimization framework developed in Chapter 7 for the occupant comfort design of wind-sensitive tall buildings. A decoupling methodology is developed to transform the originally coupled reliability-based design optimization problem into two separated sub-problems, which are then solved using the inverse reliability method and the Optimality Criteria (OC) method, respectively. The full-scale 60-story benchmark building example of mixed steel and concrete construction is employed to illustrate the effectiveness and practicality of the reliability performance-based design optimization framework.

## 8.2 Occupant perception performance function

As discussed in Chapter 5, it is the dynamic serviceability design criteria that generally govern the design of modern tall buildings. Since 1970s, much research effort has been devoted to understanding of human perception of wind-induced motion and evaluation of occupant comfort in motion of tall buildings. An extensive literature review on the development of occupant comfort design criteria is given in Chapter 5. Peak acceleration criteria associated with 10-year return period wind have been established in a number of modern design codes (NBCC 1995; JGJ 3-2002; HKCOP 2004, 2005) to preclude discomfort of occupants due to fear for safety in extreme windstorm or typhoon events. These criteria are frequency independent and are used to check acceleration performance in terms of fear for safety. Frequency dependent occupant comfort criteria in terms of 5-year-recurrence standard deviation acceleration have been established in the International Standards Organization 6897 (ISO 1984) for low frequency motion within the range of 0.063 to 1 Hz. The Architectural Institute of Japan Guidelines for the Evaluation of Habitability (AIJ-GEH 2004) established based on experimental motion simulator investigations specifies frequency dependent peak accelerations of one-year recurrence of wind, which is believed to be more closely related to frequently occurring occupant perception conditions.

As there are currently no internationally well accepted acceleration criteria which govern acceptable levels of wind-induced vibration in tall buildings, the recently proposed frequency dependent probabilistic peak modal acceleration perception guideline given in the AIJ-GEH standards (Tamura et al. 2006) are adopted in the formulation of the reliability-based design optimization framework for occupant perception performance design of tall buildings. For office buildings, one of the AIJ occupant perception criteria can be related to the ISO-6897 acceleration criteria as shown in the work by Kwok et al. (2007) using the following equation

$$a_j^U = 3.5*0.736*\exp(-3.65 - 0.41\ln f_j) \qquad (8.1)$$

where $a_j^U$ denotes the upper limits of the peak modal acceleration, corresponding to the perception probability of 90%, i.e., 90% of people in an office building may perceive the motion with the acceleration threshold given by Eq. (8.1). The value of 3.5 is a typical value of the peak factor; 0.736 is a multiplication factor, which converts the limiting acceleration from a recurrence interval of five years to one year.

Upon the determination of acceleration criterion, the successful deterministic dynamic serviceability design is ensured that the wind-induced acceleration is equal to or less than the acceleration threshold. For reliability-based design, the acceleration response should be treated as a function of some basic random variables, which represent the physical sources of variability in the evaluation of acceleration performance of a tall building. The objective of reliability-based design is to insure that the probability of the success event (acceleration response > acceleration threshold) has a satisfactory value (e.g., more than 95%). Hence, it is necessary to define an occupant perception performance function by combining the peak acceleration criterion in Eq. (8.1) and the wind-induced modal acceleration response in Eq. (5.13) as

$$
g_j(V_1, f_j, \xi_j) =
$$
$$
3.5 * 0.736 * \exp(-3.65 - 0.41 \ln f_1) - \left( \sqrt{2 \ln f_j \tau} + \gamma / \sqrt{2 \ln f_j \tau} \right) \sqrt{\frac{\pi f_j}{4 \xi_j m_j^2} S_{Q_{ij}}(f_j, V_1)} \tag{8.2}
$$

where $V_1$ denotes the design wind speed with a recurrence interval of one year; $\tau$ indicates the given time duration, and may be taken as 600s for assessing acceleration; $f_j$=the $j$-th modal frequency; $\xi_j$=the $j$-th modal damping ratio; $m_j$=the $j$-th modal mass; $S_{Q_{ij}}$=the wind-induced $j$-th modal force spectra. The dynamic serviceability failure event due to a violation in the acceleration criteria can then be defined by $g_j(V_1, f_j, \xi_j) < 0$. The multidimensional variable space can be separated into the failure domain ($g_j < 0$), the safe domain ($g_j > 0$) and the critical limit stat ($g_j = 0$). The probability of failure related to the defined limit state performance function can be generally evaluated by a multidimensional integral as follows

$$
P_f = P(g(\mathbf{x}) \leq 0) = \int_{g_j \leq 0} ... \int f_{\mathbf{x}}(\mathbf{x}) dx_1 dx_2 ... dx_n \tag{8.3}
$$

where $\mathbf{x}$ denotes the random variable vector, which is used to model uncertainty in the occupant comfort design problem; $f_{\mathbf{x}}(\mathbf{x})$ = the joint probability density function of $\mathbf{x}$.

## 8.3 Uncertainties in occupant comfort design problems

There are numerous uncertainties when designing for dynamic serviceability in tall buildings susceptible to wind-induced vibration. Uncertainties in the wind-induced vibration of tall buildings can be attributed to the random characteristics of aerodynamic wind loads or the imperfect knowledge of the structural properties of a building system. The uncertainties in wind load

194

characteristics are mainly associated with aerodynamic turbulent air flows that are inherently random. Predictions of the dynamic properties of a building may also suffer from epistemic uncertainties or modeling errors due to incomplete or inadequate information and knowledge. Therefore, it is necessary to extend the capability of the design optimization method to account for the stochastic nature of engineering design problems and thus make the existing design optimization method a more rational and effective design tool (Frangopol and Maute 2003; Schueller 2007).

Since the design wind pressure and consequently the design wind force is directly proportional to the square of the design wind speed, the determination of appropriate design wind speeds is a critical step towards the calculation of design wind loads on building structures. An accurate determination of the design wind speed for a building is generally one of the most uncertain parts of the design process (Kasperski 2007), and requires a statistical treatment on the historical data of recorded wind speeds. The estimation of design wind speeds and the corresponding estimation error (such as sampling error) has been studied in Chapter 5 as an important component in the wind-induced performance-based design optimization framework. Hereafter, the quantification of uncertainties involved in the estimation of the dynamic properties, i.e., modal frequency and modal damping ratio, of a building system is also discussed.

Since the major consideration in tall building design involved in this chapter is the occupant perception or comfort criteria, the modal frequency of a tall building becomes one of the critical system-level parameters as reflected in the occupant perception performance function defined in Eq. (8.2). The modal frequency indicating the modal stiffness of a tall building determines the spectral value $S_{Q_{jj}}(f_j, V_1)$ of the corresponding wind load and successively the resonant acceleration responses of the building. Based on some previous studies (Kareem 1988; Vrouwenvelder 2002), the lognormal distribution with a COV of 0.10 was assumed for the modal frequencies of a tall building. One reason for selecting lognormal distribution to model the uncertainty of nature frequency is that the lognormal variable can attain only nonnegative values. The mean value of the modal frequency of a tall building can be determined from a computer-based finite element model by eigenvalue analysis.

Structural damping, recognized as one of the most uncertain parameters, is another important factor in the dynamic serviceability design of tall buildings that are sensitive to wind excitation. In order to alleviate a major source of uncertainty in the design of wind-sensitive tall buildings, available data from full-scale measurements has been gathered and analyzed for the estimation of the means and the coefficient of variations (COVs) of damping values in a wide class of tall buildings (Lagomarsino 1993). The lognormal distribution is shown to provide the best fit to those measured damping data. In this study, the mean value and COV of the structural damping ratio in the first three fundamental modes are assumed to be 1% and 15% (Kareem 1988), respectively. In Table 8.1, a summary of the distribution type, statistical properties and distribution parameters for random variables are reported.

It is noted that in Table 8.1 three random variables representing the three major uncertainties involved with the occupant comfort problem have been listed. The probabilistic distribution models and the corresponding distribution parameters adopted in this study are based on the available literature, and rather than a definite way. When more and more full-scale measurement data about the dynamic properties of buildings are made available, these probabilistic distribution models and distribution parameters may be needed to update. In practice, building mass and stiffness, or underlying construction material, structural element dimension and structural form, are much less uncertain than the damping and frequency values and, therefore, can be treated as deterministic variables. For the sake of simplicity and practicality, the dimensions of structural elements are treated as deterministic design variables in the following formulation of the reliability performance-based design optimization problem.

## 8.4    Reliability Performance-based Design Optimization

Based on a wind and hurricane design framework recommended by Chock et al. (1998), multiple design wind hazard levels as shown in Table 8.2 can be explicitly defined by specific probabilities of exceedance to cover a greater spectrum of possible extreme or regularly occurring wind events for a building. In this chapter, the very frequent and occasional levels of wind events have been considered in the design of tall buildings against wind-induced vibration and drift, respectively. The very frequency wine events, such as regularly occurring synoptic gales and monsoonal winds, are believed to relate the common motion perception conditions, which can be checked against the recently recommend frequency dependent acceleration criteria (AIJ-GEH 2004). The occasional levels of wind events with 50-year return period are the commonly adopted wind load condition in modern design codes (HKCOP 2004, 2005) for wind-induced drift and the corresponding strength design.

### 8.4.1    Formulations of reliability-based structural optimization

The reliability performance-based design optimization problem of tall buildings is to be formulated with various constraints in order to satisfy different design objectives at specified performance levels. For drift serviceability design, the equivalent static wind load approach developed in Chapter 4 has been adopted with deterministic drift constraints. Actually, the wind-induced drift performance of a building always involves a deterministic static mean component and tends to have a higher level of certainty compared to the acceleration performance. The probabilistic constraints for the occupant comfort design can be formulated by introducing the target failure probability, which is the acceptable probability of occupant comfort failure events defined by $g_j < 0$.

Consider a mixed steel and concrete building having $i= 1,2, ..., N$ structural elements, including steel frame, concrete frame and shear wall elements. For simplicity, all element sizing design variables (i.e., crossing section area of steel section, dimension size of concrete section) can be represented by a collective set of generic sizing variables $z_i$, which are assembled as the deterministic design parameter vector $\mathbf{z}$. Then the minimum structural material cost design of a building structure subject to the deterministic and probabilistic performance-based constraints can be stated as:

Minimize
$$W(\mathbf{z}) = \sum_{i=1}^{N} w_i z_i \qquad (8.4)$$

Subject to

$$d_l \leq d_l^U \qquad (l = 1,2,...,N_l) \qquad (8.5)$$

$$P\{g_j(\mathbf{x},\mathbf{z}) \leq 0\} \leq P_j^U \qquad (j = 1,2,...,n) \qquad (8.6)$$

$$z_i^L \leq z_i \leq z_i^U \qquad (i= 1,2,...,N) \qquad (8.7)$$

Eq. (8.4) defines the objective function of the minimum material cost, in which $w_i$ =the respective unit cost of the steel sections, concrete sections and shear walls. Eq. (8.5) expresses the deterministic static drift performance constraints under 50-year return period wind, where $d_l^U$ denotes the design threshold value of the drift performance. In general, the allowable wind-induced drift ratio for tall buildings appears to be within the range of 1/750 to 1/250, with 1/400 being typical. Eq. (8.6) represents the set of $j = 1,2,..., n$ probabilistic constraints for dynamic serviceability performances of a tall building under the most critical incident wind angle conditions with 1-year recurrence interval wind speed, where $\mathbf{x}$ denotes the random variable vector, including three random variables, i.e., design wind speed, modal frequency and damping ratio; $P_j^U$ denotes allowable failure probability of dynamic serviceability performance of a tall building corresponding to the $j$-th mode. It is noted that the expression of performance function $g_j(\mathbf{x},\mathbf{z})$ is in a general

197

form and is an implicit function of the deterministic design parameter vector $\mathbf{z}$ as well as random variables. Eq. (8.7) defines the element sizing constraints in which superscript $L$ denotes lower size bound and superscript $U$ denotes upper size bound of member $i$.

## 8.4.2 Decoupling of stiffness optimization and probabilistic constraints

The original problem defined through Eqs. (8.4) to (8.7) can be considered as a nested optimization problem with a two-looped configuration as shown in Figure 8.1. The recursive procedure for the element sizing optimization with a minimum cost objective function is the main outer loop. The nested inner loop is aimed to treat the probabilistic constraints by performing reliability analysis and searching for the solution satisfying a prescribed target failure probability. Thus any change of the design variables in the outer loop may require for a reevaluation of the reliability of the dynamic serviceability acceleration performance, in which the reliability analysis itself is a computationally intensive numerical procedure. The numerical procedure for performing the reliability analysis can be implemented by the first-order reliability method (FORM) or the Monte Carlo Simulation method (Ditlevsen and Madsen 1996). A FORM approximation to $P_f$ is obtained by locating the most probable failure point (MPFP) on the failure surface ($g_j(\mathbf{x}, \mathbf{z}) = 0$) with minimum distance to the origin in a standard normal space. According to the work of Shinozuka (1983), the required minimum distance as a measure of reliability in the FORM method may be determined by solving an equality constrained minimization problem. The inner optimization loop for performing reliability analysis and the outer design optimization loop for searching optimum deterministic sizing design variables constitute the bi-level nested optimization problem. The direct solving of the loop-nested optimization problem is computationally expensive (Royset et al. 2001; Zou and Mahadevan 2006). In order to improve computational efficiency in the RBDO methodology, various approaches in which reliability analysis and deterministic design optimization are performed separately have been developed. Royset et al. (2001) proposed a new approach for solving the reliability-based optimal structural design problem by representing the reliability terms in traditional RBDO into deterministic functions, which define the minimum of the corresponding limit state function within a ball of specified radius. This reformulation has a limitation and cannot be applied to problems where the distribution parameters of some random variables are design variables, which is the common situation in the RBDO problems (Li and Foschi 1998; Tu et al. 1999; Lee et al. 2002; Youn et al. 2003). The approach in this study also allows the distribution parameters of random variables as design variables. Zou and Mahadevan (2006) developed an efficient methodology to solve RBDO problems by decoupling the deterministic optimization algorithm and reliability analysis procedure.

The decoupling methodology is also employed in this study by introducing an intermediate design parameter, i.e., the mean value of modal frequency $\mu_{f_j}$, in the light of the fact that the generic form of the performance function can be approximated by an explicit function of the modal frequency as shown in Eq. (8.2). By varying the mean value of modal frequency, the target reliability level of the occupant comfort performance of tall buildings may be achieved through searching for the optimum mean value of modal frequency, which makes the probabilistic acceleration performance satisfying the required target reliability index. The optimum mean value of modal frequency can be obtained with the aid of the inverse reliability method (Der Kiureghian et al. 1994; Li and Foschi 1998; Tu et al. 1999). In the design practice, it is reasonable to assume that the mean value of the modal frequency of a tall building can be determined from a computer-based finite element model by eigenvalue analysis. Under this assumption, the mean value of the modal frequency can be viewed as an implicit function of the element sizing variable $z_i$ as $\mu_{f_j}(z_i)$. Once the optimum mean value of modal frequency is available, the searching for the optimum value of the element sizing variable $z_i$ can be implemented by solving a deterministic design optimization problem subject to a set of multiple frequency and drift constraints. Hence the original reliability performance-based design optimization problem can be decoupled into two sub problems as:

Sub-problem 1: Probabilistic design optimization

Find:     the optimum mean value of modal frequency $\mu_{f_j}^*$, and the MPFP point $\mathbf{u}^*$

Subject to $\qquad\qquad P(g_j(\mathbf{x}, \mu_{f_j}(\mathbf{z})) \leq 0) \leq P_j^U$ $\qquad\qquad$ (8.8)

Sub-problem 2: Deterministic stiffness optimization

Find: $\qquad\qquad \mathbf{z} = \mathbf{z}^*$, which minimizes $W(\mathbf{z})$

Subject to $\qquad\qquad \mu_{f_j}(\mathbf{z}) \geq \mu_{f_j}^*$ $\qquad\qquad\qquad$ (8.9)

$\qquad\qquad\qquad d_l(\mathbf{z}) \leq d_l^U$ $\qquad\qquad\qquad$ (8.10)

### 8.4.3 Reliability index approach and inverse reliability method

In order to apply the FORM method, it is necessary to transform a generally non-Gaussian performance function into the equivalent one with independent normal or Gaussian variates. A general transformation for this purpose is known as the Rosenblatt transformation (Rosenblatt 1952), $\mathbf{u} = T_R(\mathbf{x})$, which transforms the original random variables into a set of standard normal variates. Then the corresponding Rosenblatt transformation of the limit-state function $g_j(\mathbf{x}, \mu_{f_j}(\mathbf{z}))$ could be defined as the following

$$G_j(\mathbf{u}, \mu_{f_j}(\mathbf{z})) = G_j\left[T_R(\mathbf{x}), \mu_{f_j}(\mathbf{z})\right] = g_j(\mathbf{x}, \mu_{f_j}(\mathbf{z})) \qquad (8.11)$$

For notation simplification, Denoting $p = \mu_{f_j}(\mathbf{z})$, Eq. (8.11) can then be rewritten as

$$G_j(\mathbf{u}, p) = g_j(\mathbf{x}, p) \qquad (8.12)$$

For generalization purpose, the design variable for the reliability analysis $p$ can also be extended to a vector by considering other kinds of distribution parameters of random variables, such as the mean value of damping ratio or the standard deviation of modal frequency. But in this study, $p$ is only considered as the mean value of modal frequency.

It should be noted that the original definition of the probabilistic dynamic serviceability constraints in Eq. (8.6) and (8.8) include two inequality relations. One represents the definition of the occupant perception failure event, the other imposes the limit on the probability of failure being within the allowable failure probability. With the aid of the reliability index approach (Tu et al. 1999; Lee et al. 2002), the two-inequality probabilistic constraints can be converted into one inequality relationship of the reliability index defined in the standard normal space. In accordance with the FORM method, the prescribed failure probability limit $P_j^U$ could be approximated in terms of a target lower bound reliability index $\beta^L$ as

$$P_j^U \approx \Phi\left(-\beta^L\right) \qquad (8.13)$$

Similarly, for the current design with a mean value of modal frequency, the probability of failure can also be related to the corresponding design reliability index as

$$P\{G_j(\mathbf{u}, p) < 0\} \approx \Phi\left(-\beta_j\right) \qquad (8.14)$$

Therefore, the original probabilistic constraints could be transformed into the form of one inequality

$$\Phi\left(-\beta_j\right) \le \Phi\left(-\beta^L\right) \quad j = 1, 2, ..., n \qquad (8.15)$$

Since the cumulative distribution function (CDF) of the standard normal distribution is a

200

non-decreasing monotonic function about its variable, the probabilistic constraints finally can be represented in terms of the reliability index as

$$\beta_j \geq \beta^L \quad j=1,2,\ldots,n \tag{8.16}$$

Utilizing the reliability index approach, the sub-problem 1 are reformulated as

Find:     optimum mean value of modal frequency $p^* = \mu^*_{f_j}$, and the MPFP point $\mathbf{u}^*$

Subject to         $\min\left(\|\mathbf{u}\| = \sqrt{\mathbf{u}^T \mathbf{u}}\right) = \beta^L$, and $G_j(\mathbf{u}, p) = 0$      (8.17)

where superscript $T$ denotes transpose operation.

The method of Lagrange's multiplier has been used to obtain the Kaush-Kuhn-Tucker optimality condition for the sub-problem 1 as (Shinozuka 1983; Der Kiureghian et al. 1994; Li and Foschi 1998)

$$\mathbf{u} + \frac{\beta^L \nabla_{\mathbf{u}} G_j}{\|\nabla_{\mathbf{u}} G_j\|} = 0 \tag{8.18}$$

$$G_j(\mathbf{u}, p) = 0 \tag{8.19}$$

where $\nabla_{\mathbf{u}} G_j$ = the gradient vector of $G_j$ with respect to $\mathbf{u}$. The modified Hasofer-Lind-Rackwitz-Fiessler (HLRF) algorithm was applied to solving the inverse reliability problem of the sub-problem 1 (Hasofer and Lind 1974; Shinozuka 1983; Liu and Kiureghian 1991) based on the following recursive formulas

$$\mathbf{u}_{k+1} = \frac{\nabla_{\mathbf{u}} G_j^T\left(\mathbf{u}_k, p_k\right)\mathbf{u}_k - G_j(\mathbf{u}_k, p_k)}{\left\|\nabla_{\mathbf{u}} G_j\left(\mathbf{u}_k, p_k\right)\right\|^2} \nabla_{\mathbf{u}} G_j\left(\mathbf{u}_k, p_k\right) \tag{8.20}$$

$$p_{k+1} = p_k + \frac{\nabla_{\mathbf{u}} G_j^T\left(\mathbf{u}_k, p_k\right)\mathbf{u}_k - G_j(\mathbf{u}_k, p_k) + \beta^L \left\|\nabla_{\mathbf{u}} G_j^T\left(\mathbf{u}_{k+1}, p_k\right)\right\|}{\left.\dfrac{\partial G_j}{\partial p}\right|_{p=p_k}} \tag{8.21}$$

where $k$ indicates the iteration number. It is worth noting that the above two recursive equations can be readily derived from the Taylor series expansions of the performance function and the KKT conditions.

The KKT condition of (8.18) and HLRF algorithm demand the gradient vector $\nabla_{\mathbf{u}} G_j$ and the partial derivative $\partial G_j / \partial p$, which will be readily available if the performance function of Eq. (8.2) is an explicit function of the random variables. Otherwise, other methods, e.g., numerical finite difference method, have to be used to quantify the gradient and derivative information of the performance function. In order to facilitate the explicit expression of the occupant comfort performance function, the modal wind force spectra $S_{Q_{jj}}(f_j, V_1)$ in Eq. (8.2) should be explicitly expressed as a function of modal frequency and design wind speed within the typical frequency range for dynamic serviceability check based on the regression analysis of normalized modal wind force spectra as (Islam et al. 1992; Chan and Chui 2006; Chan et al. 2007)

$$\frac{f_j S_{Q_{jj}}(f_j, V_1)}{\left(0.5 \rho V_1^2 HB\right)^2} = b_j \left(\frac{f_j B}{V_1}\right)^{-a_j} \tag{8.22}$$

where $H=$ the building height; $B=$ the building width; $\rho=$ the air density. Therefore, the $j$-th modal wind force spectra can be rewritten as

$$S_{Q_{jj}}(f_j, V_1) = \left(0.5 \rho V_1^2 HB\right)^2 \frac{b_j}{f_j} \left(\frac{f_j B}{V_1}\right)^{-a_j} \tag{8.23}$$

Substituting Eq. (8.23) into Eq. (8.2), the explicit occupant comfort performance function in terms of random variables $(V_1, f_j, \xi_j)$ is obtained as

$$g_j(V_1, f_j, \xi_j) =$$

$$3.5 * 0.736 * \exp(-3.65 - 0.41 \ln f_j) - \left(\sqrt{2 \ln f_j \tau} + \frac{\gamma}{\sqrt{2 \ln f_j \tau}}\right) \frac{\rho HBV_1^2}{4 m_j} \sqrt{\frac{b_j \pi}{\xi_j} \left(\frac{f_j B}{V_1}\right)^{-a_j}} \tag{8.24}$$

The three components of the gradient column vector $\nabla_x g_j = \left(\dfrac{\partial g_j}{\partial V_1}, \dfrac{\partial g_j}{\partial f_j}, \dfrac{\partial g_j}{\partial \xi_j}\right)^T$ of the performance function are given as the followings

$$\frac{\partial g_j}{\partial V_1} = -\left(\sqrt{2 \ln f_j \tau} + \frac{\gamma}{\sqrt{2 \ln f_j \tau}}\right) \left(\frac{4 + a_j}{2}\right) \frac{\rho HB}{4 m_j} \sqrt{b_j \pi B^{-a_j}} \frac{V^{1 + a_j/2}}{f_j^{a_j/2} \xi_j^{0.5}} \tag{8.25}$$

202

$$\frac{\partial g_j}{\partial f_j} = -3.5 \times \frac{0.41}{f_j} \exp(-3.65 - 0.41 \ln f_j) + \frac{\rho H B \alpha_j}{8 m_j} \sqrt{\beta_j \pi B^{-\alpha_j}} \frac{V^{2+\alpha_j/2}}{f_j^{\alpha_j/2+1} \zeta_j^{0.5}} g -$$

$$\frac{\rho H B}{4 m_j} \sqrt{\beta_j \pi B^{-\alpha_j}} \frac{V^{2+\alpha_j/2}}{f_j^{\alpha_j/2} \zeta_j^{0.5}} \left[ \frac{1}{f_j \sqrt{2 \ln f_j \tau}} - \frac{\gamma}{f_j \left(2 \ln f_j \tau\right)^{1.5}} \right]$$

(8.26)

$$\frac{\partial g_j}{\partial \zeta_j} = \left( \sqrt{2 \ln f_j \tau} + \frac{\gamma}{\sqrt{2 \ln f_j \tau}} \right) \frac{\rho H B}{8 m_j} \sqrt{\beta_j \pi B^{-\alpha_j}} \frac{V^{2+\alpha_j/2}}{f_j^{\alpha_j/2} \zeta_j^{1.5}}$$

(8.27)

Based on Rosenblatt transformation and the chain rule, the gradient vector $\nabla_{\mathbf{u}} G_j$ in the standard normal space can be related to $\nabla_{\mathbf{x}} g_j$ as

$$\nabla_{\mathbf{u}} G_j = \mathbf{J}_{\mathbf{x\_u}} \nabla_{\mathbf{x}} g_j \qquad (8.28)$$

where $\mathbf{J}_{\mathbf{x\_u}} =$ the Jacobian matrix, which is the matrix of all first-order partial derivatives with respect to $\mathbf{u}$ of a vector-valued transformation $\mathbf{x} = T_R^{-1}(\mathbf{u})$.

The partial derivative $\partial G_j / \partial p$ represents the sensitivity information of the performance function to the change of intermediate design parameter $\mu_{f_j}$. It can be related to the derivative of the performance function to modal frequency $\partial g_j / \partial f_j$ by the chain rule as

$$\frac{\partial G_j}{\partial p} = \frac{\partial G_j}{\partial f_j} \frac{\partial f_j}{\partial p} = \frac{\partial g_j}{\partial f_j} \frac{\partial f_j}{\partial p} \qquad (8.29)$$

where $\partial f_j / \partial p$ is obtained based on the transformation $f_j \rightarrow u_2$ from the original Lognormal random variable to the standard normal variate $f_j = \exp\left(u_2 \eta + \ln p - 0.5 \eta^2\right)$ as (Ang and Tang 1975)

$$\frac{\partial f_j}{\partial p} = \frac{1}{\mu_{f_j}} \exp\left(u_2 \eta + \ln \mu_{f_j} - 0.5 \eta^2\right) \qquad (8.30)$$

where $\eta =$ the standard deviation of the modal frequency's logarithm.

Once all the necessary gradient and derivative information (Eq. (8.28) and (8.30) ) are available, the iterative application of Eq. (8.20) and (8.21) to search for the optimum intermediate design parameter $p = \mu_{f_j}$ and the MPFP point satisfying a given target reliability index for the specified performance function constitutes the inverse reliability method.

### 8.4.4 Stiffness optimization subject to frequency constraints

To facilitate a numerical solution for the deterministic design optimization sub-problem 2, it is necessary that the implicit modal frequency and static drift constraints in Eqs. (8.9) to (8.10) be formulated explicitly in terms of element sizing variables $z_i$. Using the principal of virtual work, the drift responses of a tall building under ESWLs can be formulated explicitly in terms of the element sizing design variables as presented in Chapter 4.

Using the Rayleigh Quotient method, the modal frequency $f_j$ or natural period $T_j$ of a building system can be related to the total internal strain energy $U_j$ of the system due to the $j$-th modal inertia force applied statically to the system as follows (Huang and Chan, 2007):

$$T_j^2 = U_j / c_j \qquad (8.31)$$

where $c_j$ denotes a proportionality constant which relates the internal strain energy $U_j$ due to the $j$-th modal inertia loads to the square of the $j$-th modal period $T_j$ of the system. For a tall building of mixed steel frame and concrete core construction, the total internal strain energy of the building structure due to a set of externally applied modal inertia forces can be obtained by summing up the internal work done of each member as expressed in Eq. (5.50) in Chapter 5. Substituting Eq. (5.50) into Eq. (8.31), the natural period of $j$-th mode and successively the modal frequency can then be expressed as a function in terms of the element sizing design variables.

Upon establishing the explicit formulation of the modal frequency and drift design constraints, the next task is to apply a suitable numerical technique for solving the deterministic sub-problem 2 as the deterministic optimal stiffness design problem. A rigorously derived Optimality Criteria (OC) method, which has been shown to be computationally efficient for large-scale structures is herein employed (Chan 2001, Chan 2004). In this OC approach, a set of optimality criteria for the optimal design is first derived and a recursive algorithm is then applied to indirectly solve for the optimal solution by satisfying the derived optimality criteria. To seek for numerical solution using the OC method, the constrained optimal design problem must be transformed into an unconstrained Lagrangian function which involves both the objective function and the set of explicit modal frequency and drift constraints associated with corresponding Lagrangian multipliers. By differentiating the Lagragian function with respect to each sizing design variable and setting the derivatives to zero, the necessary stationary optimality conditions can be obtained and then utilized in a recursive relation to resize the active sizing variables until convergence occurs. More details of the OC method can be referred to Chapter 4.

### 8.4.5 Procedure of reliability performance-based design optimization

The overall procedure of the proposed reliability performance-based design optimization method can be outlined step by step as follows:

1. Carry out statistical analysis to estimate design wind speed corresponding to various wind hazard levels defined in Table 2.

2. Given the geometric shape of a tall building, determine the aerodynamic wind load spectra and modal force spectra by wind tunnel tests and perform regression analysis within the frequency range of serviceability design.

3. Develop the finite element model of the building and carry out an eigenvalue analysis to obtain the modal frequencies and mode shapes of the vibrations of the building.

4. Establish the explicit occupant comfort performance function in Eq. (8.24), and start the iteration procedure of the inverse reliability method with $k=0$ to search for the optimum mean value of modal frequency $p^* = \mu_{f_j}^*$.

5. For the current set of random variables $\mathbf{u}_k$, evaluate and the gradient vector using Eq. (8.28).

6. Using Eqs. (8.20) and (8.21), find the new set of random variables $\mathbf{u}_{k+1}$ and the new result of intermediate design parameter $p_{k+1}$.

7. Repeat steps 5 and 6, and check the convergence of the recursive process: if $\mathbf{u}_{k+1} = \mathbf{u}_k$ and $p_{k+1} = p_k$, then proceed to step 8 with the optimum mean value of modal frequency $p_{k+1}^* = \mu_{f_j}^*$.

8. Carry out an eigenvalue analysis and update ESWLs on the building, redo the structural analyais.

9. Establish the explicit expression of the modal frequency and drift constraints and formulate the deterministic stiffness design optimization sub-problem 2, and determine the optimal element sizing variables using the rigorously derived Optimality Criteria (OC) method.

10. Check the convergence of the design objective function: if the cost of the structure for three consecutive reanalysis-and-redesign cycles is within certain prescribed convergence criteria, for example within 0.1% difference in the structural material cost, then terminate the deterministic design optimization process to retrieve the optimized building structure with the minimum material cost and the target modal frequency, which ensures the building having a desired reliability of the occupant comfort performance; otherwise, return to step 2, update the finite element model using the current set of design variables and repeat the eigenvalue analysis and design optimization process.

The overall procedure of reliability performance-based design optimization of wind-sensitive tall buildings is also graphically presented by the flow chart as shown in Figure 8.2.

## 8.5 Illustrative example

The same 60-story benchmark building used in Chapters 5 and 6 is employed again to illustrate the effectiveness and practicality of the reliability performance-based design optimization technique. The modal force spectra are obtained from the wind tunnel test carried out at the CLP Power Wind/Wave Tunnel Facility (WWTF) of the Hong Kong University of Science and Technology (Tse et al, 2007). One-year recurrence interval wind speed in a typhoon-prone urban environment like Hong Kong is considered for calculating peak acceleration response of the building, while a 50-year return period of wind is used for predicting drift performance of the building. Two modal damping ratios for calculating the acceleration and drift responses were assumed to be 1% and 1.5%, respectively.

Due to vortex shedding effects, significant crosswind vibrations of the building in the Y-direction (short direction) induced by the 90-degree wind is found in the acceleration results presented in Chapter 5. The 90-dgree wind is then identified as a critical wind condition for the occupant comfort reliability optimization. The results of reliability assessment for the occupant comfort performance of the initial building design are presented in Table 8.3. It is found that only the reliability index associated with the first modal acceleration occupant perception constraint is less than the target reliability index of 2, which is generally adopted for the serviceability design (Ang and Tang1975; Wen 2001). The small value of reliability index means that the occupant comfort performance function of the first mode violates the corresponding probabilistic acceleration constraint. Therefore, the inverse reliability method is employed to search for the optimum mean value of the first modal frequency, by which the target reliability index is achieved. The iteration history of the optimum mean value of the first modal frequency is given in Table 8.4. After 10 iterations, the intermediate design parameter is found to be 0.2357 Hz, which satisfies the peak acceleration criteria with a target reliability of 2. The optimized mean value of first modal frequency has been increased by 27% when compared with the initial modal frequency value of 0.185 Hz.

For comparison, the reliability assessment of occupant comfort performance for the optimized building design is performed using both FORM and Monte Carlo simulations. It is found in the final reliability results shown in Table 8.5 that the reliability index for the first vibration mode is very

206

close to the target value of 2. Such a result verifies the accuracy of the inverse reliability method in searching for the optimum mean value of modal frequency satisfying the specified target reliability index. The reliability indexes for both the second and third modes are found to be well above the target reliability index. Actually, due to the improvement in the overall building lateral and torsioanl stiffnesses of the building, the reliability indexes for both the second and third modes are increased by 8.4% and 43.9% respectively. The larger increase in the occupant comfort reliability index for the torsional vibration mode 3 indicates that the torsional stiffness has been much more enhanced by the deterministic element sizing optimization process.

Figure 8.3 presents the material cost design history of the building in the deterministic stiffness design optimization subject to modal frequency and drift constraints. The normalized cost with respect to the initial cost of the building is given for each design cycle, which includes the process of one formal structural analysis and one deterministic resizing optimization. Two design history curves are presented in Figure 8.3. One cost history curve is obtained using the developed modal frequency and drift optimization technique, while the other is obtained for comparison purpose by using the conventional static drift optimization method without considering the modal frequency constraint. Although the structural costs of the building are found somewhat fluctuating at the first few design cycles, steady convergence to the final optimum solution is achieved at the seventh design cycle for the drift optimization only and the eighth design cycle for the modal frequency and drift optimization. If only optimizing drift performances, the performance-based optimization technique is able to achieve about 4% saving in the material cost. When the modal frequency constraints is taken into account in the design optimization process, a moderate increase of about 7% in the structural cost is needed to fulfill the first modal frequency constraint in addition to the multiple static drift constraints. An increase in the consumption of structural cost is expected as the reliability of dynamic serviceability of the optimized building has been improved.

Figure 8.4 presents the design history of the first there modal frequencies. A steady and rapid convergence of the first modal frequency to the target mean first modal frequency value of 0.2357 Hz obtained from the sub-problem 1 has been achieved for the large-scale 60-story building structure. As demonstrated in the example, the OC method is capable of searching for the optimal distribution of element stiffness of practical building structures to satisfy simultaneously modal frequency and static drift design constraints. By successfully solving the inverse reliability problem and the deterministic stiffness design optimization problem, the developed reliability performance-based structural optimization is able to deliver a most cost-effective design solution while satisfying the probabilistic occupant perception constraints and the deterministic drift constraints.

207

## 8.6 Summary

This chapter presents a decoupling methodology for reliability performance-based structural optimization of wind-sensitive tall building designs. The occupant comfort performance function for the dynamic serviceability design of a tall building is expressed explicitly in terms of three random variables, i.e., design wind speed, modal frequency and damping ratio, which represent the major uncertainties of wind characteristics and system properties. The original two-loop coupled reliability-based design optimization (RBDO) problem is decomposed into two separated sub-problems, in terms of an inverse reliability problem and a deterministic stiffness optimization problem. These two sub-problems are then effectively solved using the inverse reliability method and Optimality Criteria (OC) method, respectively.

A traditional RBDO methodology is very computationally expensive for practical large engineering systems. To overcome the computational difficulty, a new unilevel and sequential formulation of RBDO is developed. The KKT conditions corresponding to the probabilistic constraint for occupant comfort design have been derived analytically and utilized for searching the optimum mean value of modal frequency and the MPFP satisfying the specified target reliability index corresponding to the most active probabilistic performance constraint. The deterministic stiffness design optimization augmented by the inverse reliability formulation is mathematically equivalent to the original nested reliability-based optimization formulation if the reliability optimization problem is solved by satisfying the KKT conditions. The inverse reliability method is illustrated using the 60-story benchmark building, and has been found to be numerically robust and computationally efficient. The accuracy of the inverse reliability method is also verified by the forward FORM method and the Monte Carlo simulation.

As shown in the design history results of the 60-story benchmark building, the most cost efficient design solution is obtained within only a few number of reanalysis and redesign cycles. The developed reliability performance-based design optimization technique provides structural engineers a new tool for the probabilistic dynamic serviceability performance design of tall buildings under various levels of wind loads.

**Table 8.1    Random variables in wind-induced occupant comfort problem**

| Random variables | Distribution model | Statistical properties | | |
| --- | --- | --- | --- | --- |
| | | Mean | Standard deviation | COV |
| Wind speed (m/s) | Gumbel | 22.5 | 1.125 | 0.05 |
| Natural frequency (Hz) | Lognormal | 0.185 | 0.0185 | 0.1 |
| Damping ratio | Lognormal | 0.01 | 0.0015 | 0.15 |

**Table 8.2    Performance-based wind hazard design level**

| Wind design severity level | Design wind speed[*] (m/s) | Average return period | Probability of exceedance | Performance levels |
| --- | --- | --- | --- | --- |
| Very Frequent | 22.5 | 1 year | 100% in 50 years | Occupant perception |
| Frequect | 30.6 | 5 years | 99.9% in 50 years | Occupant comfort |
| Frequent | 34.7 | 10 years | 99.5% in 50 years | Fear for safety |
| Occasional | 46.9 | 50 years | 64% in 50 years | Drift / Strength |
| Rare | 55.7 | 475 years | 10% in 50 years | Safety |

[*]Design wind speed is referred at the height of 90 m in Hong Kong area.

**Table 8.3    The reliability index and probability of failure for the initial building**

| Mode | Mean frequency (Hz) | FORM | | Monte Carlo Simulation | |
| --- | --- | --- | --- | --- | --- |
| | | Reliability index | Probability of failure | Probability of failure | Sample number |
| Mode 1 | 0.185 | 1.163 | 1.224E-01 | 1.158E-01 | 1000 |
| Mode 2 | 0.327 | 6.483 | 4.504E-11 | 0 | 500,000 |
| Mode 3 | 0.410 | 2.641 | 4.128E-03 | 4.006E-03 | 50,000 |

**Table 8.4    Iteration history using inverse reliability method**

| Iteration number | Random vector: $\mathbf{u}_k$ | | | $\mu_{f_1}^*$ | $\|\mathbf{u}_k\|$ | $G_k(\mathbf{u}, p)$ |
|---|---|---|---|---|---|---|
| 1 | 0.0746 | 0.1773 | 0.0499 | 0.185 | 0.1987 | 0.0281 |
| 2 | -0.3385 | 0.9171 | -0.3904 | 0.1975 | 1.0526 | 0.0112 |
| 3 | -0.4821 | 1.3308 | -0.5711 | 0.2134 | 1.5263 | 0.0052 |
| 4 | -0.5231 | 1.5598 | -0.6326 | 0.2243 | 1.7626 | 0.0029 |
| 5 | -0.5281 | 1.6851 | -0.6472 | 0.2302 | 1.8808 | 0.0017 |
| 6 | -0.5227 | 1.7531 | -0.6462 | 0.2332 | 1.9401 | 0.001 |
| 7 | -0.5158 | 1.7898 | -0.6413 | 0.2346 | 1.9699 | 0.0006 |
| 8 | -0.5101 | 1.8095 | -0.6366 | 0.2352 | 1.9849 | 0.0003 |
| 9 | -0.5021 | 1.8307 | -0.6296 | 0.2358 | 1.9953 | 0 |
| 10 | -0.4996 | 1.8323 | -0.6271 | 0.2357 | 1.9996 | 0 |

**Table 8.5    The reliability index and probability of failure for the optimized building**

| Mode | Mean frequency (Hz) | FORM | | Monte Carlo Simulation | |
|---|---|---|---|---|---|
| | | Reliability index | Probability of failure | Probability of failure | Sample number |
| Mode 1 | 0.236 | 1.999 | 2.276E-02 | 2.41E-02 | 1000 |
| Mode 2 | 0.496 | 7.028 | 1.050E-12 | 0 | 500,000 |
| Mode 3 | 0.589 | 3.801 | 7.206E-05 | 8.0E-05 | 200,000 |

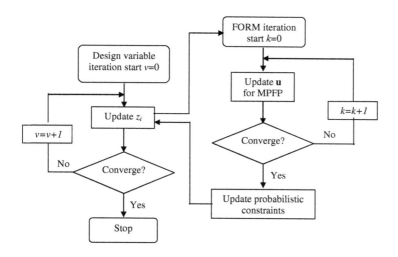

**Figure 8.1  Two-loop nested configuration of reliability-based structural optimization**

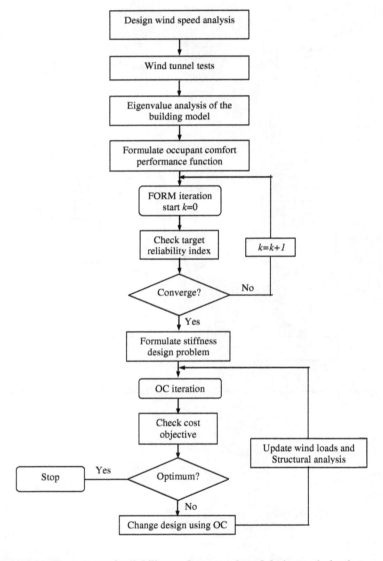

**Figure 8.2 Flow chart of reliability performance-based design optimization**

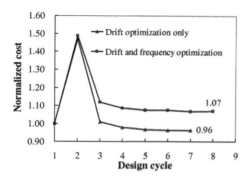

**Figure 8.3  Design history of structure cost of the 60-story building**

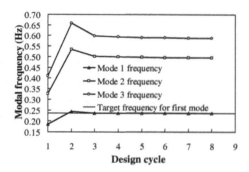

**Figure 8.4  Design history of modal frequencies of the 60-story building**

213

# CHAPTER 9   Conclusions and Recommendations

## 9.1   Conclusions

The major effort of this research is the development of a general design optimization methodology for performance-based design of tall building structures that are sensitive and susceptible to wind loads. Specifically, the reliability performance-based optimization method accounting for the stochastic nature of winds and tall buildings has been developed for seeking the most cost-effective stiffness design of a tall building subject to both static drift and dynamic serviceability design criteria.

In this research, the analytical method for the prediction of wind-induced dynamic response of a tall building has been investigated firstly. The integrated wind load determination and stiffness design optimization of tall buildings subject to static drift criteria is then developed. To provide an effective and efficient computational tool for solving the dynamic serviceability or occupant comfort design problem of modern tall buildings, the performance-based design optimization technique is proposed for the mitigation of wind-induced vibrations in tall buildings. Since wind excitations and thus wind-induced vibrations are inherently random, the deterministic optimization method is extended to the reliability-based probabilistic structural optimization approach taking into account the major uncertainties arising from either aerodynamic wind loading characteristics or building system properties.

The time domain approach is employed to investigate the inherent variability of random vibration in terms of peak factors for the prediction of expected largest peak response of a tall building. The time-variant reliability analysis of wind-excited tall buildings is developed and integrated into the reliability-based performance design optimization method, which is able to achieve an optimal stiffness design solution satisfying both deterministic and probabilistic drift constraints. The epistemic uncertainties, which have a direct impact on the occupant comfort evaluation of a tall building, are identified and quantified in order to facilitate a general framework for reliability performance-based design optimization of a tall building subject to probabilistic occupant comfort design constraints. The reliability performance-based design optimization methodology is successfully developed based on the inverse reliability method and the Optimality Criteria method.

Based on the analytical development and numerical results presented, the following conclusions on the wind-induced performance-based design optimization of tall buildings are drawn.

## 1. Coupled dynamic analysis

The general coupled dynamic analysis method of tall buildings subject to wind excitations is developed based on random vibration theory. The cross-correlation of modal responses of tall buildings under spatiotemporally varying dynamic wind loads has been investigated in detail. Three different formulae for determining the intermodal correlation coefficient with various levels of accuracy have been established. Corresponding to these three different formulae for the intermodal correlation, three modal combination methods, namely the traditional CQC (TCQC), the accurate CQC (ACQC) and the exact CQC (ECQC) methods, have been developed for predicting wind-induced dynamic responses of tall buildings. It is found in the numerical results that the statistical coupling among modal responses in terms of intermodal correlation coefficient does have a significant influence on the total dynamic acceleration response of a tall building with closely spaced modal frequencies and 3D mode shapes.

## 2. Integrated wind load updating and stiffness design optimization

After developing the procedure to determine the equivalent static wind loads on tall buildings, the dependence of wind-induced loads on natural frequencies of tall buildings is revealed and the necessity of wind load updating during design optimization process is highlighted. It is found that the dependency of wind-induced loads on natural frequencies is more significant in the crosswind and torsional wind directions due to resonant effects than in the alongwind directions. By stiffening the building structure and updating wind-induced structural loads during the optimization process, the benefit of wind load reduction especially in the crosswind or torsional direction can be fully utilized in the integrated wind load updating and stiffness design optimization technique. The most cost efficient stiffness distribution of the building structure satisfying multiple interstory drift design criteria has been achieved as demonstrated in two tall building examples. The practicality and effectiveness of the integrated design optimization method have been illustrated by a 40-sotry public housing building design subject to practical design constraints.

## 3. Performance-based dynamic serviceability design optimization

Since the different performance levels of a tall building design can be related to the design wind speed in association with the specific recurrence intervals, the extreme value analysis is proposed to estimate performance-based design wind speed in a local site, in which the building is situated. The sampling error due to the use of a limited number of recorded typhoon wind speed data in the estimation of design wind speeds is also quantified. It is believed that a more accurate prediction of site-specific extreme wind loads can often lead to more cost efficient designs of tall buildings. The evaluation of occupant comfort performance of a tall building requires checking on wind-induced vibration, which can be evaluated in terms of modal acceleration and peak resultant acceleration using experimentally derived aerodynamic wind load spectra and a semi-analytical technique based on random vibration theory. It is worth noting that the peak resultant acceleration resulted from the modal combination and torsional amplification effects generally represents the most critical acceleration design constraint in a tall building design.

The optimal dynamic serviceability design of a tall building subject to peak resultant acceleration or modal acceleration criteria is then formulated explicitly in terms of element sizing variables and solved by the rigorously derived Optimality Criteria method. The stiffness optimization method for the mitigation of wind-induced vibrations in tall buildings provide a powerful design tool to deliver a permanent and most economical stiffness design solution for a tall building with satisfactory drift and occupant comfort performance. Both frequency independent peak resultant acceleration criteria and frequency dependent standard deviation or peak modal acceleration criteria have been considered in the optimization process. It has been found that various occupant comfort acceleration criteria have different implication on the evaluation of habitability and the final optimal design solution does depend on the particular choice of occupant comfort criteria.

## 4. Peak response statistics and reliability of tall buildings

As demonstrated in Chapter 6, the probabilistic distribution of the largest value of a random response process has a complementary relationship to the first-passage time distribution. Based on the classical solution to the first-passage problem in random vibration theory (Rice 1945; Vanmarcke 1975), the probabilistic peak factor has been derived and explicitly expressed as a function of excitation duration, the probability of exceedance and the bandwidth of a random process of interest. For a Gaussian scalar process, the Davenport peak factor can be considered as a special case of the probabilistic peak factor for a wide band process with a specific exceedance probability of 43%. The conservativeness of the Davenport peak factor for a narrow band process is also demonstrated. The procedure to assess time-variant reliability of wind-induced motions in a tall building has been established using the information of peak response probabilistic distributions.

A generalized peak factor formula is derived from the Weibull distribution model using asymptotic theory of statistical extremes. The proposed Weibull peak factor has a shape parameter, which can be determined by statistical treatment on the time history response process of a tall building. Such a statistical-based method to predict the expected largest peak response in terms of the Weibull peak factors takes full utilization of the time history wind-induced response results. It is found that using the statistical-based method to predict peak component acceleration can attain the same level of reliability as using the conventional method based on Davenport's peak factor.

Furthermore, the Gamma peak factor has been obtained to facilitate the prediction of the expected maximum value of a resultant response process composed of two orthogonal component processes. Based on the Gamma peak factor, the combined process method is developed for predicting analytically the expected maximum resultant response of a wind-excited tall building without the need of time history analysis. The reliability implication of various methods in the estimation of expected maximum dynamic response is also studied.

## 5. Mathematical formulation of reliability-based performance design optimization framework

A reliability-based performance design optimization framework integrated with peak value statistical analysis method is mathematically formulated for the optimal stiffness design of a tall building under random wind excitation. The primary difficulty of the engineering optimization problem is the treatment of numerous and various constraints through the whole design process, especially the time and spatial depending constraints. The developed optimization procedure also takes into account the inherent uncertainty in wind-induced random vibrations by estimating the expected largest peak response and assessing time-variant reliability of tall buildings against wind-induced motion.

## 6. Reliability performance-based design optimization

The dynamic serviceability design limit state of tall buildings is firstly established to facilitate formulating reliability-based structural optimization problem by defining the occupant perception performance function, which is expressed in terms of random variables representing the major uncertainties in the characteristics of wind and the properties of a building system. The original coupled two-loop reliability-based design optimization problem is reformulated and decoupled into two separated sub-problems, i.e., the inverse reliability problem and the deterministic stiffness design optimization problem. These two sub-problems are then effectively solved using the inverse reliability method and the Optimality Criteria (OC) method, respectively. The inverse reliability problem is defined to seek for the optimum mean value of modal frequency satisfying a specified target reliability index corresponding to the occupant perception design limit state. Once the optimum mean modal frequency value is available, the deterministic stiffness design optimization is invoked to obtain the optimum stiffness design solution of a tall building satisfying the deterministic drift and probabilistic occupant comfort performance design constraints.

The developed inverse reliability procedure based on FORM method allows the direct and efficient determination of the intermediate design parameter, the mean value of modal frequency, which makes the probabilistic occupant perception acceleration performance satisfying the required target reliability index. The efficiency and robustness of the procedure has been verified by a forward FORM method and Monte Carlo simulation using a full-scale 60-story benchmark building example. Given the desired mean value of modal frequency, the OC method is employed to solve the optimal stiffness design of the benchmark building subject to frequency and drift constraints. The encouraging results of the benchmark building have shown that the developed reliability performance-based deign optimization method provides a useful and efficient design tool to optimize the tall building design with satisfactory and reliable drift and occupant comfort performance under various levels of wind excitations.

## 7. Effective and efficient Optimality Criteria (OC) method

The rigorously derived OC technique developed in this research is effective and computationally efficient. Not only is it applicable for static and deterministic structural design optimization, it is also suitable for wind-induced dynamic and probabilistic response optimization of tall buildings. The optimal design method does not depend on the initial design and the number of design variables, but rather on the structural behaviors of tall buildings.

The rapid and steady convergence of the OC method for stiffness optimization can be always achieved as demonstrated by a number of academic examples and practical applications presented in relevant chapters. Such a satisfactory convergence performance of the numerical OC algorithm could be attributed to the virtual work based convex formulation and the particular structural behavior of a tall building, which overall behaves like a cantilever structure under lateral loads and its internal force field is generally insensitive to changes in member sizes.

## 9.2   Recommendations for future work

During the course of this research, several issues concerning the related findings, methods and a review of literature arose, indicating needs for future developments in both theory and applications. The following recommendations are made for future research.

# 1. Nonlinear coupled dynamic analysis

As discussed in Section 2.2, dynamic response analysis plays a central role in the design course of structural systems. For normal wind-resistant design of tall buildings, it is reasonable and safe to assume the building systems are linear. In this research, only the frequency and occasional recurrence levels of wind hazards have been formerly treated by the performance-based optimization technique with the assumption that under these wind events structural responses are in the linear-elastic stage. It is noted that in the recommended performance-based wind engineering design level in Table 2.1 the rare extreme wind of 475-year or even 1000-year return period may cause nonlinear-inelastic responses of building structures. Therefore, nonlinear dynamic analysis or an inelastic "pushover" analysis has to be used in the future development to cover the design against the rare extreme wind hazard.

As reviewed in section of 2.2.1, the dynamic analysis for a randomly excited nonlinear multi-degree-of-freedom (MDOF) system is one of the most difficult problems in the random vibration field. Zhu et al. (1990, 1996, 2001) proposed a general functional form and solution method of the exact stationary solution for MDOF nonlinear systems with Gaussian white noise excitations using the concepts of integrability, resonance, and the property of Poisson bracket in Hamiltonian dynamics. Based on these analytical solutions, it is interesting to develop a general equivalent design method, rather than "pushover" method, for nonlinear design and optimization of tall building under extreme wind or earthquake excitations.

# 2. Simultaneous consideration of wind, earthquake and gravity effects

The research has been limited to the stiffness design optimization of a tall building under wind loadings. As discussed in Chapter 2, seismic effects on tall buildings may be quite different from wind effects due to the different sources and characteristics of the two loading mechanisms. Response spectrum method (Der Kiureghian 1981) is widely adopted for seismic design of building structures. Since the earthquake response spectrum ordinate increases as the modal frequency of a building increases, the stiffness optimization may be contradictory to that of wind-resistant design, in which the wind load spectral value generally decreases with an increase in the modal frequency. The new performance-based optimization approach may find a balance stiffness design point satisfying the demands from both earthquake and wind. Furthermore, in terms of load resistant system, design for gravity loads may be quite different from the design for lateral loads. Therefore, the optimal performance-based design technique needs further exploration for simultaneously considerations of all effects under wind, earthquake and gravity loadings.

# 3. Development and application of two-dimensional wind velocity (magnitude and direction) climate model

Probability distribution of extreme wind speeds can be evaluated using observed or simulated extreme wind speed data. Since severe wind events are statistically rare events, a simulation method is required to produce enough wind speed data to facilitate statistical extreme value analysis. The extreme value distribution analysis is to be conducted on each set of directional wind speed data to estimate the joint probability density function of wind speed and direction. Since the past extreme wind events in Hong Kong were mainly based on tropical cyclones, the focus of the climate simulation may be placed on the simulation of tropical cyclones. Tropical cyclones can be described using Newton's second law augmented by conservation laws for mass, thermodynamic energy, and water vapor. One of the most rigorously developed wind climate models is the MM5 model (Grell et al. 1994), which has been developed as an open source code.

From the numerical simulation of the probability distribution of extreme wind speeds, it is possible to obtain the directionality of wind and integrate it into the directionality of dynamic response and thereby estimate the extreme value distribution of dynamic response. The closed form expression of extreme value distribution of dynamic response can be used to directly assess probability for reliability-based design. In addition, such a closed form distribution can be effectively used to formulate and simplify probabilistic design constraints.

The probabilistic characteristic of extreme wind events can be described by estimated joint probability density function of wind speed and direction. The relationship between the extreme wind events and the response of the building system can be established using the random vibration method. The annual probability (or the mean annual frequency) that the wind-induced response exceeds any specified threshold value, can be formulated using the total probability theorem (Cornell et al. 2002). To facilitate the computation, the probability of interest can be expanded by conditioning on all possible levels of directional extreme wind events (say, 5yr-return-period North wind, 10yr-return-period South wind and 50yr-return-period Eastern wind).

# 4. Statistics of multivariate extremes and first-passage problem for a vector stochastic process

The out-crossing probability describing the probability that at least one of the component processes exceed its respective threshold is essential for estimating the reliability of a structural system whose response is a vector stochastic process. For wind-induced response of a tall building, in Chapter 6, the out-crossing probability of a vector stochastic process has been reduced to the consideration of one critical process, which happens at a critical position, i.e., at the top level or at the weakest story. However, for any other structural systems, it may not easy to identify the so-called "critical position". A general and more accurate method to estimate the out-crossing probability of a vector stochastic process is therefore needed, e.g., using simulation approach (Au and Beck 2001). Although a closed form solution, the so-called generalized Rice formula, for multi-dimensional first passage problem is available (Belyaev 1968), the solution involves with some multi-dimensional integrals, which require generally extensive computational effort (Song and Der Kiureghian 2006).

Using the asymptotic approach for statistical extremes, Davenport (1964) obtained an analytical peak factor formula for a Gaussian process, which is widely used in wind engineering. As demonstrated in Chapter 6, the Weibull peak factor is derived from the asymptotic theory of statistical extremes for any given stochastic process. It may also be possible to solve the first-passage problem of a vector stochastic process using statistics of multivariate extremes, which itself is being developed by applied mathematician and seems quite promising to have a wide application in the engineering field (Kotz and Nadarajah 2000; Gupta and Manohar 2005).

## 5. Time history analysis and optimal acceleration design of tall buildings

Dynamic serviceability acceleration design optimization technique for wind-sensitive tall buildings has been developed with the aid of spectral analysis and presented in Chapter 5 and 8.　In Chapter 6, the dynamic response optimization method is investigated in the time domain for the lateral drift design of a tall building structure. It is also possible to further develop an optimal acceleration design technique in the time domain. The maximum instantaneous acceleration at a critical time instant can be related to the displacement response using the time-domain finite-difference method, i.e., the central difference method. Once the critical acceleration response is approximated by the difference of the corresponding displacement response points in the time axis discretized using a small time step, the explicit expression of maximum acceleration can be obtained and then the OC technique can be applied to search for the optimal structural system while satisfying the acceleration design constraints.

## 6. Mathematical programming approach with equilibrium constraints for inverse reliability problem

The inverse reliability problem posted in Chapter 6 is a general optimization problem with one equality constraint indicating the limit-sate surface. The problem is solved successfully by the method of Lagrange's multiplier partly due to the explicitly expression of the limit-state surface for occupant comfort design. In general, the limit-sate function in the design of engineering system is implicitly governed by ordinary or partial differential equations (ODEs or PDEs). Therefore, the method of Lagrange's multiplier becomes difficult to apply. Mathematical programming with equilibrium constraints (MPECs) deals with the problem in which the essential constraints are defined by parametric variational inequality to model physical equilibrium phenomena (Luo et al. 1996). As a general optimization approach without the limit of explicit expression, mathematical programming with equilibrium constraints is a choice to further develop a general and powerful computer-based tool for the reliability-based design optimization.

## 7. Integrated performance-based optimal design of stiffness, vibration control and health monitoring of building structures

Vibration control concepts and technologies have been developed for civil engineering structures to reduced excessive vibrations cased by strong winds, severe earthquakes or other excitations (Housner et al. 1997; Spencer et al. 2003). Due to the high initial investment and life-cycle maintenance cost of control systems, the application of structural control technology is limited. Recently, there has been a renewed interest in measuring the dynamic properties of the building structures in terms of performance monitoring (Xu et al. 2000; Kwok 2004), or in detecting possible damage of the infrastructure after an extreme event or long-term service in terms of health monitoring (Aktan et al. 2000). The information gathered by performance or health monitoring is found to be useful to incorporate into an existing or planned structural control scheme (Gattulli and Romeo 2000). Very recently, Xu and Chen (2007a and 2007b) proposed a methodology for integrated vibration control and health monitoring of building structures using semi-active friction dampers. The integrated approach seems more practical and cost-effective since the building structure only needs one system to serve both for controlling and monitoring.

It is noted that the vibration control and health monitoring idea is consistent with the performance-based design concept to design buildings that have a predictable and reliable performance under harsh environments. Therefore, the planning and design of the integrated control and monitoring system should be implemented during the building design stage under the uniform framework of the optimal performance-based design technique. In the future optimal performance-based design platform, not only the stiffness, but also mass and damping become adjustable through the determination of the sensory system, data acquisition and transmission system, and the selection of proper system identification, damage detection and control algorithms. The proposed integration can be firstly implemented by introducing an overall design objective cost function including not only the structural material cost, but also the initial investment and life-cycle operational and maintenance cost of control and monitoring systems. The performance indexes of building structures with and without control can be considered as the major design constraints.

## 8. Probabilistic objective function

In this study, a relatively simple and deterministic structural material cost function is taken as the design objective function. It is necessary to find a better objective function that reflects more accurately the material cost, construction cost and the expected damage-induced repair cost. In order to estimate damage-induced repair cost, a comprehensive damage loss model due to failure events needs to be developed (Cheng et al. 1998). Since the future damage is essentially probabilistic, the failure probability can be included in the definition of design objective function in the reliability-based design optimization framework (Royset et al. 2001). Much more challenges are associated with solving optimal structural design problems involving the failure probability in both the objective and constraint functions than that of the problems involving only with probabilistic constraints (Royset et al. 2006).

# Reference

Ad Hoc Committee on Serviceability Research. "Structural serviceability: a critical appraisal and research needs." *Journal of Structural Engineering*, ASCE, 1986, 112, 2646-2664.

Ang, A. H. S., and Tang, W. H. (1975). Probability concepts in engineering planning and design, John Wiley, Vols I and II.

Aktan AE, Catbas FN, Grimmelsman KA, Tsikos CJ. (2000). "Issues in infrastructure health monitoring for management." *Journal of Engineering Mechanics*, ASCE, 126(7), 711–724.

Allam S.M. and Datta T.K. (1999). "Seismic behavior of cable-stayed bridges under multi-component random ground motion." *Engineering Structures*, 21, 62-74.

Allam S.M. and Datta T.K. (2004). "Analysis of cable-stayed bridges under multi-component random ground motion by response spectrum method" *Earthquake Engineering and Structural Dynamics*, 33, 375-393.

Applied Technology Council (ATC). ( 1997). "NEHRP guidelines, commentary and example applications for the seismic rehabilitation of buildings (FEMA 273)." ATC-33, Redwood City, Calif.

Architectural Institute of Japan Recommendations (2004). Guidelines for the evaluation of habitability to building vibration, AIJ-GEH-2004, Tokyo, Japan.

Arora JS, Cardoso JEB (1989). "A design sensitivity analysis principle and its implementation into ADINA." *Computers and Structures*, 32, 691–705.

Arora JS (1995). "Structural design sensitivity analysis: continuum and discrete approaches." In: Herskovits J (ed) Advances in structural optimization, 47–70, Kluwer Academic, Boston

Arora and Wang, Q. (2005). "Review of formulations for structural and mechanical system optimization." *Structural and Multidisciplinary Optimization*, 30, 251–272.

ASEC 7-98. (1999). "Standard minimum design loads for buildings and other structures." American Society of Civil Engineers, Reston VA.

Au, S.K., and Beck, J.L. (2001). "First excursion probabilities for linear systems by very efficient importance sampling." *Probabilistic Engineering Mechanics*, 16, 193-207.

Banavalkar, P.V. (1990). "Structural systems to improve wind induced dynamic performance." *Journal of Wind Engineering and Industrial Aerodynamics* 36, 213–224.
Belyaev, Y. K. (1968). "On the number of exits across the boundary of a region by a vector stochastic process." *Theory of Probability and its Applications*, 13(2), 320–324.

Bhatti, M.A., and Pister, K.S. (1981). "A dual criteria approach for optimal design of earthquake-resistant structural systems." *Earthquake Engineering and Structural Dynamics*, 9, 557-572.

Bleistein, N., and Handelsman, R. (1986). Asymptotic expansions of integrals. Dover, New York, N.Y.

Boggs, D., (1995). "Acceleration indexes for human comfort in tall buildings-peak or RMS?" CTBUH Monograph Chapter 13, Council on Tall Buildings and Urban Habitat.

Boggs D.W., Hosoya N., Cochran L. (2000). "Sources of torsional wind loading on tall buildings: lessons from the wind tunnel." In: Proceedings of the 2000 Structures Congress & Exposition.

Bogomolni M., Kirsch U., and Sheinman I. (2006). "Efficient design sensitivities of structures subjected to dynamic loading." *International Journal of Solids and Structures*, 43, 5485–5500.

Breitung, K. (1984). "Asymptotic approximations for multinormal integrals." *Journal of Engineering Mechanics*, ASCE, 110(3), 357-366.

Breitung, K. (1991). "Probability approximations by log likelihood maximization." *Journal of Engineering Mechanics*, 117. 457-477.

Burton, M. D., Kwok, K. C. S., Hitchcock, P. A. and Roy, O.D. (2006). "Frequency dependence of human response to wind-induced building motion." *Journal of Structural Engineering*, 132(2), 296-303.

Burton, M. D., Kwok, K. C. S. and Hitchcock, P. A. (2007). "Occupant comfort criteria for wind-excited buildings: based on motion duration." *Proc. 12th International Conference on Wind Engineering*, Cairns, Australia, 2 – 6 July, 2007, 1207-1214.

Cai GQ, Lin YK. (1988). "A new approximate solution technique for randomly excited non-linear oscillators." *International Journal of Non-Linear Mechanics*, 23, 409–420.

Carassale, L., and Solar, G. (2006). "Monte Carlo simulation of wind velocity fields on complex structures." *Journal of Wind Engineering and Industrial Aerodynamics*, 94, 323–339.

Caughey, T.K. (1971). "Nonlinear theory of random vibrations." Advances in Applied Mechanics, 11, Academic Press, New York, 209-253.

Caughey, T.K., and Ma, F. (1982). "The exact steady-state solution of a class of nonlinear stochastic systems." *International Journal of Non-Linear Mechanics*, 17(3), 137–142.

Cermak, J.E. (1977). "Wind-tunnel testing of structures." *Journal of the Engineering Mechanics Division*, ASCE 103, 1125–1140.

Cermak, J.E. (2003). "Wind-tunnel development and trends in applications to civil engineering." *Journal of Wind Engineering and Industrial Aerodynamics*, 91, 355–370.

Chan, C.M. 1992. "An optimality criteria algorithm for tall steel building design using commercial standard sections." *Structural Optimization*, 5, 26–29.

Chan, C.M, Grierson DE, Sherbourne AN. 1995. "Automatic optimal design of tall steel building frameworks." *Journal of Structural Engineering*, ASCE, 121(5), 838–847.

Chan, C.M. (1997). "How to optimize tall steel building frameworks." In ASCE Manuals and Report on Engineering Practice, No. 90, Guide to Structural Optimization, J. Arora (ed.). ASCE,

165–195.

Chan, C.M. (1998). "Optimal stiffness design to limit static and dynamic wind responses of tall steel buildings." *Engineering Journal*, American Institute of Steel Construction, Third Quarter, 94–105.

Chan, C. M. (2001). "Optimal lateral stiffness design of tall buildings of mixed steel and concrete construction." *Journal of Structural Design of Tall Buildings* 10(3), 155-177.

Chan, C. M. (2004). "Advances in structural optimization of tall buildings in Hong Kong." *Proc. 3rd China-Japan-Korea Joint Symposium on Optimization of Structural and Mechanical Systems*, Kanazawa, Japan, 30 Oct. – 2 Nov., 2004, 49-57.

Chan, C.M. and Zou X.K. (2004). "Elastic and inelastic drift performance optimization for reinforced concrete building under earthquake loads." *Earthquake Engineering and Structural Dynamics*, 33(8), 929-950.

Chan C.M. and Chui J.K.L. (2006). "Wind-induced response and serviceability design optimization of tall steel buildings." *Engineering Structures*, 28(4), 503-513.

Chan C.M., Chui J.K.L. and Huang M.F. (2007). "Integrated aerodynamic load determination and stiffness optimization of tall buildings." In press in *Journal of Structural Design of Tall and Special Buildings* (published online in July 2007).

Chan C.M., and Wong K.M. (2007). "Structural topology and element sizing design optimization of tall steel frameworks using a hybrid OCGA method." *Journal of Structural and Multidisciplinary Optimization* (accepted for publication).

Chandu, S.V.L., and Grandhi, R.V. (1995). "General purpose procedure for reliability based structural optimization under parametric uncertainties." *Advances in Engineering Software*, 23, Elsevier Science Limited, 7-14.

Chang C.C., Ger J.F. and Cheng F.Y. (1994). "Reliability-based optimum design for UBC and nondeterministic seismic spectra." *Journal of Structural of Engineering*, 120(1), 139-160.

Chang, F.K. (1967). "Wind and movement in tall buildings." *Civil Engineering*, ASCE, 37(8), 70-72.

Chang, F.K. (1973). "Human response to motions in tall buildings." *Journal of Structural Division*, ASCE, 99, 1259-1272.

Chen, L., and Letchford, C. W., (2004a). "A deterministic-stochastic hybrid model of downbursts and its impact on a cantilevered structure." Engineering Structure, 26(5), 619-629.

Chen, L., and Letchford, C. W., (2004b). "Parametric study on the along-wind response of the CAARC building to downbursts in the time domain." *Journal of Wind Engineering and Industrial Aerodynamics*, 92(9), 703-724.

Chen, P.W., and Robertson, L.E. (1973). "Human perception threshold of horizontal motion." *Journal of Structural Division*, ASCE, 98(8), 1681-1695.

Chen X, Kareem A. (2004). "Equivalent static wind loads on tall buildings: New model." *Journal of*

*Structural Engineering*, ASCE, 130(10), 1425-1435.

Chen, X, and Kareem, A. (2005a). "Coupled dynamic analysis and equivalent static wind loads on buildings with three-dimensional modes." *Journal of Structural Engineering*, 131(7), 1071-1082.

Chen, X, and Kareem, A. (2005b). "Dynamic wind effects on buildings with 3D coupled Modes: Application of high frequency force balance measurements." *Journal of Engineering Mechanics*, 131(11), 1115-1125.

Chen, X. (2007). "Prediction of Alongwind Tall Building Response to Transient Winds."*Proc. 12th International Conference on Wind Engineering*, Cairns, Australia, 2 – 6 July, 2007, 263-270.

Cheng, F.Y., and Truman, K.Z. (1983). "Optimization algorithm of 3D building systems for static and seismic loading." Modeling and Simulation in Engineering, W.F. Ames, ed., North-Holland Pub. Co., 315-326.

Cheng, G., Li G., and Cai Y. (1998). "Reliability-based structural optimization under hazard loads". *Structural Optimization*, 16, 128-135.

Cheng, P.W., Bussel, G.J.W., Kuik, G.A.M., and Vugts, J.H. (2003). "Reliability-based design methods to determine the extreme response distribution of offshore wind turbines." *Wind Energy*, 6(1), 1-22.

Chock, G., Boggs, D. and Peterka, J. (1998). "A wind and hurricane design framework for multi-hazard performance-based engineering of high-rise buildings". Structural Engineering World Wide, T139-3.

Choi KK, Haug EJ, Seong HG (1983). "An iterative method for finite dimensional structural optimization problems with repeated eigenvalues." *International Journal for Numerical Methods in Engineering*, 19, 93–112.

Choi, K.K., and Kim, N.H. (2005). Structural sensitivity analysis and optimization. Springer Science, New York.

Choi, WS, and Park, GJ. (2002). "Structural optimization using equivalent static loads at all time intervals." *Computer Methods in Applied Mechanics and Engineering*, 191, 2077–2094.

Chopra A.K. (2000). Dynamics of Structures: theory and applications to earthquake engineering. Prentice-Hall, New Jersey.

Clark. T.L., Teddie Keller., Janice Coen., Peter Neilley., Hsiao-Ming Hsu and William D. Hall. (1997). "Terrain-Induced Turbulence over Lantau Island of Hong Kong: 7 June 1994 Tropical Storm Russ Case Study." *Journal of the Atmospheric Sciences*, 54, 1795-1814.

Clough R.W. and Penzien J. (1993). Dynamics of Structures, McGraw-Hill, New York.

Cornell C.A., Jalayer F., Hamburger R.O., and Foutch D.A. (2002). "Probabilistic basis for 2000 SAC Federal Emergency Management Agency steel moment frame guidelines." *Journal of Structural Engineering*, ASCE, 128 (4), 526-533.

Cramer H. (1946). Mathematical Methods of Statistics. Princeton University Press, Princeton.

Crandall, S.H. (1963). "Zero crossings, peaks, and other statistical measures of random responses." *Journal of the Acoustical Society of America*, 35(11), 1693-1699.

Davenport, A. G. (1961). "The spectrum of horizontal gustiness near the ground in high winds." *Quarterly Journal of the Royal Meteorological Society*, 88(376), 194-211.

Davenport, A. G. (1963). "The buffeting of structures by gusts." Proceedings, International Conference on Wind Effects on Buildings and Structures, Teddington U.K., 26–8 June, 358–391.

Davenport, A. G. (1964). "Note on the distribution of the largest value of a random function with application to gust loading." *Proceedings, Institution of Civil Engineering*, 28, 187-196.

Davenport, A. G., and Isyumov, N. (1967). "The application of the boundary layer wind tunnel to the prediction of wind loading." Proceedings, International Research Seminar on Wind Effects on Buildings and Structures, Ottawa, Canada, 11–15 September, 201–230.

Davenport AG. (1995)."How can we simplify and generalize wind loading?" *Journal of Wind Engineering and Industrial Aerodynamics*, 54/55, 657-669.

Deodatis, G. (1996). "Simulation of ergodic multivariate stochastic processes." Journal of Engineering Mechanics, 112(8), 778-787.

Der Kiureghian, A. (1980). "Structural response to stationary excitation." *Journal of Engineering Mechanics*, 106(6), 1195-1213.

Der Kiureghian, A. (1981). "A response spectrum method for random vibration analysis of MDF system." *Earthquake Engineering and Structural Dynamics*, 9, 419-435.

Der Kiureghian., A., Lin., H.Z., and Hwang, S.J. (1987). "Second-order reliability approximations." *Journal of Engineering Mechanics*, 113, 1208-1225.

Der Kiureghian, A., and Nakamura, Y. (1993). "CQC modal combination rule for high-frequency modes." Earthquake Engineering and Structural Dynamics, 22, 943-956.

Der Kiureghian, A., Zhang, Y., Li, C.C. (1994). "Inverse reliability problem." *Journal of Structural Engineering*, 120, 1154-1159.

Der Kiureghian., A. (1996). "Structural reliability methods for seismic safety assessment: a review." *Engineering Structures*, 18(6), 412-42.

Der Kiureghian., A. (2000). "The geometry of random vibrations and solutions by FORM and SORM." *Probabilistic Engineering Mechanics,* 15, 81–90

Di Paola, M., and Muscolino, G. (1990). "Differential moment equations of FE modeled structures with geometrical non-linearities." *International Journal of Non-Linear Mechanics*, 25(4), 363-373.

Di Paola, M., Falsone, G., and Pirrotta, A. (1992). "Stochastic response analysis of nonlinear systems under Gaussian inputs." *Probabilistic Engineering Mechanics*, 7, 15-21.

Dimentberg, M.F. (1982). "An exact solution to a certain nonlinear random vibration problem." *International Journal of Non-Linear Mechanics*, 17(4), 231-236.

Dimentberg, M.F. (2005). "Random vibrations of a rotating shaft with non-linear damping." *International Journal of Non-Linear Mechanics*, 40, 711 – 713.

Er, G. K. (1998). "Multi-Gaussian closure method for randomly excited nonlinearsystems." *International Journal of Non-Linear Mechanics*, 33, 201–14.

Enevoldsen, I., and Sorensen, J.D. (1994). "Reliability-based optimization in structural engineering." *Structural Safety*, 15, 169-196.

Evgrafov A, Patriksson M (2003). "Stochastic structural topology optimization: discretization and penalty function approach." *Structural and Multidisciplinary Optimization*, 25(3):174–188

Federal Emergency Management Agency (FEMA). (1997). "NEHRP Guidelines and commentary for Seismic Rehabilitation of Buildings." FEMA-273, 1997.

Federal Emergency Management Agency (FEMA). (2000a). "Action plan for performance based seimic design." FEMA-349, SAC Joint Venture, Washington, D.C.

Federal Emergency Management Agency (FEMA). (2000b). "Recommended seismic design criteria for new steel moment frame buildings." FEMA-350, SAC Joint Venture Washington, D.C.

Ferris MC, and Pang, JS. (1997). "Engineering and economic applications of complementarity problems." *Society for Industrial and Applied Mathematics Review*, 39(4), 669–713.

Ferris MC, and Tin-Loi F (2001). "Limit analysis of frictional block assemblies as a mathematical program with complementarity constraints." *International Journal of Mechanical Sciences*, 43(1), 209–224.

Fiacco A.V., McCormick G.P. (1990). Nonlinear Programming: Squential Unconstrained Minimization Techniques. SIAM, Philadelphia.

Fishman G.S. (1996). Monte Carlo: concepts algorithms, and applications. Springer series in operations research, New York: Springer.

Flay, G.J., Buttgereit, V.O., Bailey, K.I., Obasaju, E., and Brendling, W.J. (2003). "A comparison of force and pressure measurements on a tall building with coupled mode shapes." *Proceedings of the 11th International Conference on Wind Engineering*, 2-5 June 2003, Texas Tech University, 2373-2380.

Fletcher, R., and Powell, M.J.D. (1963). "A rapidly convergent descent method for minimization." The Computer Journal, 6, 163-168.

Fletcher, R., and Reeves, C.M. (1964). "Function minimization by conjugate gradients." The Computer Journal, 7, 149-154.

Foley, C. M. (2002). "Optimized performance-based design for buildings". Recent Advance in Optimal Structural Design. Burns SA (ed.), ASCE, Reston, VA. 169-240.

Frangopol, D.M. (1985a). "Sensitivity of reliability-based optimum design." *Journal of Structural Engineering*, ASCE, 111(8), 1703-1721.

Frangopol, D.M. (1985b), "Structural optimization using reliability concepts design." *Journal of*

*Structural Engineering*, ASCE, 111(11), 2288-2301.

Frangopol, D.M., and Maute, K. (2003). "Life-cycle reliability-based optimization of civil and aerospace structures." *Computers and Structures*, 81, 397-410.

Gan C.B., Zhu W.Q. (2001). "First-passage failure of quasi non-integrable-Hamiltonian systems." *International Journal of Non-Linear Mechanics*, 36, 209-220.

Ganzerli, S, Pantelides, C. P., and Reaveley, L. D. (2000). "Performance-based design using structural optimization." *Earthquake Engineering and Structural Dynamics*, 29, 1677-1690.

Gattulli V, Romeo F. (2000). "Integrated procedure for identification and control of MDOF structures." *Journal of Engineering Mechanics*, ASCE, 126(7), 730–737.

Glanville, M.J., Kwok, K.C.S., and Denoon, R.O. (1996). "Full-scale damping measurements of structures in Australia." *Journal of Wind engineering and Industrial Aerodynamics*, 59, 349-364.

Goldenberg, Stanley B., Christopher W. Landsea, et al. (2001). "The Recent Increase in Atlantic Hurricane Activity: Causes and Implications." Science, 293(July 20), 474-479.

Gong, Y.L., Xu, L., and Grierson, D.E. (2005). "Performance-based design sensitivity analysis of steel moment frames under earthquake loading." *International Journal for Numerical Methods in Engineering*, 63, 1229–1249.

Gomes, L., and Vickery, B.J. (1976). "On the prediction of extreme wind speeds from the parent distribution." *Journal of Wind engineering and Industrial Aerodynamics*, 2(1)

Goto, T., Iwasa, Y., and Tsurumaki, H. (1990). "An experimental study on the relationship between motion and habitability in a tall residential building." *Proceedings of Tall Buildings: 2000 and Beyond, Fourth World Congress*, Hong Kong, 817-829.

Grandall S.H. (1963). "Zero crossings, peaks, and other statistical measures of random responses." Journal of the Acoustical Society of America, 35(11), 1693-1699.

Greene WH, Haftka RT (1991). "Computational aspects of sensitivity calculations in linear transient structural analysis." *Structural Optimization*, 3, 176–201.

Grell G. A., Dudhia J. and Stauffer DR. (1994). "A description of the fifth-generation Penn State/NCAR mesoscale model (MM5)." NCAR Technical Note, NCAR/TN-398+STR.

Griffis, L. G. (1993). "Serviceability limit states under wind load." AISC *Engineering Journal* 30(1), 1-16.

Grigoriu, M. (1982). "Estimates of design wind from short records." Journal of Structural Division, 108(ST5), 1034-1048.

Gumbel E.J. (1958). Statistical of Extremes. Columbia Univ. Press., New York.

Guo K. (1999). "A consistent method for the solution to reduced FPK equation in statistical mechanics." *Physica* A 262, 118-128.

Gupta, S., and Manohar C.S. "Multivariate extreme value distributions for random vibration

Applications." *Journal of Engineering Mechanics*, 131(7), 712-720.

Gurley, K.R., Tognarelli, M.A., and Kareem, A. (1997). "Analysis and simulation tools for wind engineering." *Probabilistic Engineering Mechanics*, 12(1), 9-31.

Haftka, R.T., and Gurdal, Z. (1992). Elements of structural optimization. Kluwer Academic Publishers, Dordrecht.

Hansen, R.J., Reed, J.W., and Vanmarcke, E.H. (1973). "Human response to wind-induced motion of buildings." *Journal of Structural Division*, ASCE, 99(ST7), 1589-1605.

Harbitz, A. (1986). "An efficient sampling method for probability of failure calculation." *Structural Safety*, 3, 109-115.

Harris R. I. (1963). "The response of structures to gusts." Proceedings, International Conference on Wind Effects on Buildings and Structures, Teddington U.K., 26–8 June, 394–421.

Harris, R. I. (1968) 'On the spectrum and auto-correlation function of gustiness in high winds', Electrical Research Association. Report 5273.

Hasofer, A.M., and Lind, N. (1974). "An exact and invariant first-order reliability format." *Journal of Engineering Mechanics*, ASCE, 100(EM1), 111-121.

Haug EJ, Arora JS (1979). "Applied optimal design: mechanical and structural systems." Wiley-Interscience, New York

Heredia_Zavoni E. and Vanmarcke E.H. (1994). " Seismic random-vibration analysis of multisupport-structural systems.", *Journal of Engineering Mechanics*, 120(5), 1107-1128.

Hilding D, Klarbring A, Petersson J. (1999). "Optimization of structures in unilateral contact." *Applied Mechanics Reviews*, 52(4):139–160

Holmes, J. D., and Oliver, S. E., (2000). "An empirical model of a downburst." *Engineering Structure*, 22, 1167-1172.

Holmes J.D. (2002). "Effective static load distributions in wind engineering." *Journal of Wind Engineering and Industrial Aerodynamics*, 90, 91-109.

Holemes, J.D. and Cochran L.S. (2003). "Probability distributions of extreme pressure coefficients." *Journal of Wind Engineering and Industrial Aerodynamics*, 91, 893-901.

Holmes, J.D., Rofail, A., and Aurelius, L. (2003). "High-frequency base balance methodologies for tall buildings with torsional and coupled resonant modes." Proceedings of the 11th International Conference on Wind Engineering, 2-5 June 2003, Texas Tech University, 2381-2387.

Holmes, J., Forristall, G., and McConochie, J., (2005). "Dynamic response of structures to thunderstorm winds. "Proceedings of 10th Americas Conference on Wind Engineering, Baton Rouge, Louisiana, USA

Hong Kong Code of Practice (2004). "Code of Practice for Structural Use of Concrete." Buildings Department, Hong Kong.

Hong Kong Code of Practice (2005). "Code of Practice for Structural Use of Steel." Buildings Department, Hong Kong.

Hong H.P., Beadle S., and Escobar J.A. (2001). Probabilistic assessment of wind-sensitive structures with uncertain parameters. *Journal of Wind Engineering and Industrial Aerodynamics*, 89, 893-910.

Housner GW, Bergman LA, Caughey TK, Chassiakos AG, Claus RO, Masri SF, et al. (1997) "Structural control: past present, and future." *Journal of Engineering Mechanics,* ASCE, 123:897–971.

Householder, A.S. (2006). Principles of numerical analysis. Dover Publications, New York.

Hsieh, C.C. and Arora, J.S. (1984). "Design sensitivity analysis and optimization of dynamic response." *Computer Methods in Applied Mechanics and Engineering*, 43, 195-219.

Hsieh, C.C. and Arora, J.S. (1985). "A hybrid formulation for treatment of point-wise state variable constraints in dynamic response optimization." *Computer Methods in Applied Mechanics and Engineering*, 48, 171-189.

Hsieh, C.C. and Arora, J.S. (1986). "Algorithms for point-wise state variable constraints in structural optimization." *Computer and Structures*, 22(3), 225-238.

Huang, M. F., and Chan, C. M. (2007). "Sensitivity analysis of multi-story steel building frameworks under wind and earthquake loading." *Proc. 7th World Congresses of Structural and Multidisciplinary Optimization*, COEX Seoul, Korea, 21 – 25 May, 2007, 1376-1385.

Huang, M. F., Chan, C. M., Kwok, K. C. S., and Hitchcock, P. A. (2007). "Dynamic analysis of wind-induced lateral-torsional response of tall buildings with coupled modes." *Proc. 12th International Conference on Wind Engineering*, Cairns, Australia, 2 – 6 July, 2007, 295-302.

Huang, M. F., Chan, C. M., Kwok, K. C. S., and Hitchcock, P. A. (2008). "Cross correlations of modal responses of tall buildings in wind-induced lateral-torsional motion". Submitted to *Journal of Engineering Mechanics*, ASCE.

Ibrahim, R.A., Soundararajan, A., and Heo, H. (1985). "Stochastic response of non-linear dynamic systems based on a non- Gaussian Closure." *Journal of Applied Mechanics*, ASME, 52(4), 965-970.

International Organization for Standardization (1984). "Guidelines for the Evaluation of the Response of Occupants of Fixed Structures, Especially Buildings and Offshore Structures, to Low-Frequency Horizontal Motion (0.063 to 1.0 Hz)." ISO 6897:1984, International Organization for Standardization, Geneva, Switzerland.

Irwin, A.W. (1978). "Human response to dynamic motion of structures." *The structural engineer*, 56(9), 237-243.

Irwin, P.A., and Baker, W.F. (2005). "The wind engineering of the Burj Dubai." Proc. 7th World Congress of the Council on Tall Buildings and Urban Habitat, New York, 17-19 October.

Irwin, P.A. (2006). "Developing wind engineering techniques to optimize design and reduce risk." Proc. 7th UK Conference on Wind Engineering, Wind Engineering Society, ICE.

Islam, S., Ellingwood, B., and Corotis, R. B. (1992). "Wind-induced response of structurally asymmetric high-rise buildings." *Journal of Structural Engineering*, 118(1), 207-222.

Isyumov, N., Mascantonio, A., and Davenport, A. G. (1988). "Measured building motions of tall buildings in wind and their evaluation." Symposium/Workshop on Serviceability of Buildings (Movements, Deformations, Vibrations), Ottawa, Canada, 16-18 May.

Isyumov, N., Fediw, A.A., Colaco, J. and Banavalkar, P.V. (1992). "Performance of a tall building under wind action." *Journal of Wind Engineering and Industrial Aerodynamics*, 41-44, 1053-1064.

Isyumov, N. (1994). "Criteria for acceptable wind-induced motions." *Proceedings of the Structural Congress*, ASCE, Atlanta, USA, 24-28 April, 642-647.

Isyumov, N., and Kilpatrick, J. (1996). "Full-scale experience with wind-induced motions of tall buildings." *Proceedings of the 67th Regional Conference Council on Tall Buildings and Urban Habitat*, Chicago, US, 15-18 April, 401-411.

Jain A., Spinivasan M., and Hart G. C., (2001). "Performance based design extreme wind loads on a tall building." *The structural design of tall buildings*, 10, 9-26.

Jensen, H.A., and Sepulveda, A.E. (1998). "Design sensitivity metric for structural dynamic response." *AIAA Journal*, 36, 1686–1693.

Jensen, H.A. (2005). "Structural optimization of linear dynamical systems under stochastic excitation: a moving reliability database approach." *Computer Methods in Applied Mechanics and Engineering*, 194, 1757–1778.

Kaimal, J.C. et al. (1972). "Spectral characteristics of surface-layer turbulence." *Journal of the Royal Meteorological Society*, 98, 563-589.

Kang BS, Choi WS, Park GJ (2001). "Structural optimization under equivalent static loads transformed from dynamic loads based on displacement." *Computer and Structure*, 79, 145–154.

Kang, BS, Park GJ, and Arora JS (2006). "A review of optimization of structures subjected to transient loads." *Structural and Multidisciplinary Optimization*, 31, 81–95.

Kareem, A. (1985). "Lateral-torsional motion of tall buildings to wind loads." *Journal of Structural Engineering*, 111(11), 2479-2496.

Kareem, A. (1987). "Wind effects on structures: a probabilistic viewpoint." *Probabilistic Engineering Mechanics*, 2(4), 166-200.

Kareem, A. (1988). "Aerodynamic response of structures with parametric uncertainties." Structural Safety, 5, 205-225.

Kareem, A. (1992). "Dynamic response of high-rise buildings to stochastic wind loads." *Journal of Wind engineering and Industrial Aerodynamics*, 41-44, 1101-1112.

Kareem A., and Gurley K. (1996). "Damping in structures: its evaluation and treatment of uncertainty." *Journal of Wind Engineering and Industrial Aerodynamics*, 59, 131-157.

Kareem, A., Kabat, S., Haan Jr., F.L. (1998). "Aerodynamics of Nanjing Tower: A case study."

*Journal of Wind Engineering and Industrial Aerodynamics*, 77-78, 725-739.

Kareem A, Zhou Y. (2003). "Gust loading factor-past, present and future." *Journal of Wind Engineering and Industrial Aerodynamics*, 91, 1301-1328.

Kasperski, M. (2007). "Specification of the design wind load: a critical review of code concepts." Proc. 12th International Conference on Wind Engineering, Cairns, Australia, 2 – 6 July, 2007, 41-83.

Katafygiotis, L.S., and Beck, J.L. (1995). "A very efficient moment calculation method for uncertain linear dynamic systems." *Probabilistic Engineering Mechanics*, 10, 117-128.

Katafygiotis, L.S., and Cheung, S.H. (2004). "Wedge simulation method for calculating the reliability of linear dynamical systems." *Probabilistic Engineering Mechanics*, 19, 229–238.

Katafygiotis, L.S., and Cheung, S.H. (2006). "Domain Decomposition Method for Calculating the Failure Probability of Linear Dynamic Systems Subjected to Gaussian Stochastic Loads." *Journal of Engineering Mechanics*, 132(5). 475-486.

Kim, C., and Choi, K.K. (2007). "Reliability-based design optimization using response surface method considering prediction interval estimation." *Proc. 7th World Congresses of Structural and Multidisciplinary Optimization*, COEX Seoul, Korea, 21 – 25 May, 1184-1193.

Kim, S.H., and Wen, Y.K. (1990). "Optimization of structures under stochastic loads." *Structural Safety*, 7(2-4), 177-190.

Kim, T., and Foutch, D.A. (2007). "Application of FEMA methodology to RC shear wall buildings governed by flexure." *Engineering Structures*, 29, 2514–2522.

Kirsch, U. (1993). Structural Optimization. Springer-Verlag, Berlin.

Kirsch, U. (2000). "Combined approximations – a general approach for structural optimization." *Structural and Multidisciplinary Optimization*, 20, 97–106

Kirsch U. (2003). "Design-oriented analysis of structures- a unified approach." *Journal of Engineering Mechanics*, 129, 264-272.

Kirsch U., Bogomolni M., and Sheinman I. (2007). "Efficient dynamic reanalysis of structures." *Journal of Structural Engineering*, 133(3), 440-448.

Kirsch U., and Bogomolni M. (2007). "Nonlinear and dynamic structural analysis using combined approximations." *Computers and Structures*, 85, 566–578.

Kitamura, H., Y. Tamura and T. Ohkuma (1995). "Wind resistant design and response control in Japan. Part III: structural damping and response control." Proc. 5th World Congress on Habitat and High-Rise Buildings, Tradition and Innovation, Amsterdam, The Netherlands,

Krishnamurti, T.N., D. Oosterhof, and N. Dignon, (1989). "Hurricane prediction with a high-resolution global model." *Monthly Weather Review*, 117, 631-669.

Kocer FY, Arora JS (2002). "Optimal design of latticed towers subjected to earthquake loading." *Journal of Structural Engineering*, 128, 197–204.

Kotz, S., and Nadarajah, S. (2000). "Extreme value distributions." Imperial College Press, London.

Kumar, D., and Datta, T.K. (2008). "Stochastic response of articulated leg platform in probability domain", *Probabilistic Engineering Mechanics*, in press.

Kurtaran H, Eskandarian A, Marzougui D, Bedewi NE (2002). "Crashworthiness design optimization using successive response surface approximations." *Computational mechanics*, 29, 409–421.

Kwok, K.C.S. (2004). "Wind effects on tall buildings in typhoon-prone urban environment." *Proc. 1st International Symposium on Wind Effects on Buildings and Urban Environment*, Tokyo.

Kwok, K. C. S., Burton, M. D., and Hitchcock, P. A. (2007). "Occupant comfort and perception of motion in wind-excited tall buildings." *Proc. 12th International Conference on Wind Engineering*, Cairns, Australia, 2 – 6 July, 2007, 101-115.

Lagomarsino, S. (1993). "Forecast models for damping and vibration periods of buildings." *Journal of Wind Engineering and Industrial Aerodynamics*, 48, 221-239.

Lagomarsino, S. and Pagnini L.C. (1995). "Criteria for modelling and predicting dynamic parameters of buildings." Report ISC-II, 1, Istituto di Scienza delle Construzioni, University of Genova, Genova, Italy.

Lee, B. E. (1983). "The perception of the wind-induced vibration of a tall building-a personal viewpoint." *Journal of Wind Engineering and Industrial Aerodynamics*, 12, 379-384.

Lee, J.O., Yang, Y.S., and Ruy, W.S. (2002). "A comparative study on reliability-index and target-performance-based probabilistic structural design optimization." *Computers and Structures*, 80, 257-269.

Lee TH (1999). "An adjoint variable method for structural design sensitivity analysis of a distinct eigenvalue problem." *Korean Society of Mechanical Engineers Intenational Journal*, 13, 470–479.

Letchford, C. W., Mans, C., and Chay, M. T. (2001). "Thunderstorms-their importance in Wind Engineering, a case for the next generation wind tunnel." *Journal of Wind Engineering and Industrial Aerodynamics*, 89, 31-43.

Li, H., and Foschi, R.O. (1998). "An inverse reliability method and its application." *Structural Safety*, 20, 257-270.

Li, J., and Chen, J.B. (2006). "The probability density evolution method for dynamic response analysis of non-linear stochastic structures." *International Journal for Numerical Methods in Engineering*, 65, 882–903.

Liang, J.W., Chaudhuri, S.R., and Shinozuka, M. (2007). "Simulation of nonstationary stochastic processes by spectral representation." Journal of Engineering Mechanics, 133(6), 616-627.

Liu, P.L., and Der Kiureghian. A. (1990). "Optimization algorithms for structural reliability." *Structural Safety*, 9, 161-177.

Lin Y.K. (1967). Probabilistic theory of structural dynamics. Mcgraw-Hill, New York.

Lin, Y.K. and Cai, G.Q. (1995). Probabilistic Structural Dynamics: Advanced Theory and Applications. McGraw Hill, New York.

Lin, Y.K., and Cai, G.Q. (2000). "Some thoughts on averaging techniques in stochastic dynamics." *Probabilistic Engineering Mechanics*, 15,7–14.

Liu Q, Davies HG. (1990). "The non-stationary response probability density functions of non-linearly damped oscillators subjected to white noise excitations." *Journal of Sound Vibration*, 139(3), 425–435.

Liu Y, Zhang DL, and Yau MK, (1997). "A multi-scale numerical study of hurricane Andrew, 1992: Part I: Explicit simulation and verification." *Monthly Weather Review*, 125, 3073-3093.

Liu Y, Zhang DL, and Yau MK, (1999). "A multi-scale numerical study of hurricane Andrew, 1992: Part II: Kinematics and inner-core structures." *Monthly Weather Review*, 125, 3073-3093.

Luo, Z.Q., Pang, J.S., Ralph, D. (1996). "Mathematical Programs with Equilibrium Constraints." Cambridge University Press, Cambridge

Matsui M., Ishihara T., and Hibi K. (2002). "Directional characteristics of probability distribution of extreme wind speeds by typhoon simulation." *Journal of Wind Engineering and Industrial Aerodynamics*. 90 1541-1553.

McCullough, M., and Kareem, A. (2007). "Global warming and hurricane intensity and frequency: The debate continues." *Proc. 12th International Conference on Wind Engineering*, Cairns, Australia, 2 – 6 July, 2007, 647-654.

Meirovitch L. (1986). Elements of vibration analysis. McGraw-Hill, New York.

Melbourne W.H. (1977). "Probability distributions associated with the wind loading of structures." Civil Engineering Transactions, Institution of Engineers, Australia, 19, 58-67.

Melbourne WH. (1980). "Comparison of measurements on the CAARC standard tall building model in simulated model wind flows." *Journal of Wind Engineering and Industrial Aerodynamics*, 6, 73-88.

Melbourne, W. H., and Palmer, T. R. (1992). "Accelerations and comfort criteria for buildings undergong complex motions." *Journal of Wind Engineering and Industrial Aerodynamics*, 41-44, 105-116.

Middleton, D. (1960). An introduction to statistical communication theory. McGraw-Hill, New York.

Minciarelli F., Gioffre M., Grigoriu M. and Simiu E. (2001). "Estimates of extreme wind effects and wind load factors: influence of knowledge uncertainties". *Probabilistic Engineering Mechanics*, 16, 331-340.

Montgomery DC (2001). Design and analysis of experiments. 5th edn. Wiley, Massachusetts.

Moses, F. (1969). "Approach to structural reliability and optimization." An Introduction to Structural Optimization, Cohn, M.Z. (ed.), Solid Mechanics Division, University of Waterloo,

236

Study No. 1, 81-120.

Murakami, S. (2002). "Setting the scene: CFD and symposium overview." *Wind and Structures*, 5(2-4), 83-88.

Muscolino, G., and Palmeri, A. (2005). "Maximum response statistics of MDoF linear structures excited by non-stationary random processes." *Computer Methods in Applied Mechanics and Engineering*, 194, 1711–1737.

National Building Code of Canada (1977). "Structural Commentaries (Part 4)." National Research Council of Canada, Ottawa, Ontario.

National Building Code of Canada (1995). "Structural Commentaries (Part 4)." National Research Council of Canada, Ottawa, Ontario.

National Standard of the People's Republic of China (2001). "Code for seismic design of buildings (GB 50011-2001)." Beijing, China.

National Standard of the People's Republic of China (2002). "Technical Specification for Concrete Structures of Tall Building (JGJ 3-2002)." New World Press: Beijing, China.

Newland D.E. (1984). Random vibration and spectral analysis. Longman Scientific & Technical, UK.

Newmark, N.M. (1959). "A method of computation for structural dynamics." *Journal of Engineering Mechanics Division*, ASCE, 85, 67-94.

Nigam N.C. (1983). Introduction to random vibrations. The MIT Press, Massachusetts

Ooyama, K., (1969). "Numerical simulation of the lifecycle of tropical cyclones." *Journal of Atmospheric Sciences*, 26, 3-40.

Pagnini, L.C., and Solari G. (1998). "Serviceability criteria for wind-induced acceleration and damping uncertainties." *Journal of Wind Engineering and Industrial Aerodynamics*, 74-76, 1067-1078.

Panagiotis, A.M., and Christopher, G.P. (2002). "Weight minimization of displacement-constrained truss structures using a strain energy criterion." *Computer Methods in Applied Mechanics and Engineering*, 191, 2159–2177.

Pantelides, C.P. (1990). "Optimum design of actively controlled structures." *Earthquake Engineering and Structural Dynamics*, 19(4), 583-596.

Paola, M.D., and Sofi, A. (2002). "Approximate solution of the Fokker–Planck–Kolmogorov equation." *Probabilistic Engineering Mechanics*, 17, 369–384.

Papadimitriou, C., and Lutes, L.D. (1996). "Stochastic cumulant analysis of MDOF systems with polynomial-type nonlinearities." *Probabilistic Engineering Mechanics*, 11, 1-13.

Papadimitriou, C., Katafygiotis, L.S., and Lutes L.D. (1999). "Response cumulants of nonlinear systems subject to external and multiplicative excitations." *Probabilistic Engineering Mechanics*, 14, 149-160.

237

Payne, M.C., Teter, M.P., Allan, D.C., Arias, T.A., and Joannopoulos, J.D. (1992) "Iterative minimization techniques for ab initio total-energy calculations: molecular dynamics and conjugate gradient." Reviews of Modern Physics, 64(4), 1045-1097.

Pedersen, P., and Pedersen, N.L. (2005). "An optimality criterion for shape optimization in eigenfrequency problems." *Structural and multidisciplinary optimization*, 29, 457-469.

Peterka, J.A. (1992). "Improved extreme wind predictions in the United States." *Journal of Wind engineering and Industrial Aerodynamics*, 41, 433-541.

Piterbarg, V. I. (1996). Asymptotic methods in the theory of Gaussian processes and fields. AMS, Transl. Math. Monographs, 148.

Prager, W. and Taylor, J.E. (1968). "Problems of optimal structural design." *Journal of Applied Mechanics*, 35, 102-106.

Priestley, M.B. (1965). "Evolutionary spectra and non-stationary processes." *Journal of the Royal Statistical Society*, Series B, 27, 204-237.

Powell, M.J.D. (1964). "An efficient method of finding the minimum of a function of several variables without calculating derivatives." *The Computer Journal*, 7, 155-162.

Proppe, C. (2003). "Exact stationary probability density functions for non-linear systems under Poisson white noise excitation." *International Journal of Non-Linear Mechanics*, 38, 557–564.

Proppe, C., Pradlwarter, H.J. Schueller, G.I. (2003). "Equivalent linearization and Monte Carlo simulation in stochastic dynamics." *Probabilistic Engineering Mechanics*, 18, 1–15.

Rackwitz, R. (2001). "Reliability analysis: a review and some perspectives." Structural Safety, 23, 365-395.

Razani, R. (1965). "The behaviour of the fully stressed design of structures and its relationship to minimum weight design." *AIAA Journal*, 3, 2262-2268.

Reinhold, T. (ed.) (1982). "Wind tunnel modeling for civil engineering applications." International Workshop on Wind Tunnel Modeling Criteria and Techniques in Civil Engineering Applications, Gaithersburg, Maryland, U.S.A.: Cambridge University Press.

Rice, S.O. (1945). "Mathematical analysis of random noise." *Bell System Technical Journal*, 24, 46-156.

Robert, J.A., and Hasofer, A.M. (1976). "Level crossings for random fields." *The Annals of Probability*, 4(1), 1-12.

Roberts, J.B., and Spanos, P.D. (1986). "Stochastic averaging: an approximate method of solving random vibration problems." *International Journal of Non-Linear Mechanics*, 21, 111–34.

Roberts, J.B., and Spanos, P.D. (1990). Random vibration and statistical linearization. Wiley, New York.

Rosenblatt, M. (1952). "Remarks on a multivariate transformation." *The Annals of Mathematical Statistics*, 23(3), 470-472.

Royset, J. O., Der Kiureghian, A., and Polak, E. (2001). "Reliability-based optimal design of series structural systems." *Journal of Engineering Mechanics*, 127(6), 607–614.

Royset, J.O., Der Kiureghian, A., and Polak, E. (2006). "Optimal Design with Probabilistic Objective and Constraints." *Journal of Engineering Mechanics*, 132(1), 107-118.

Schmit, L.A. (1960). "Structural design by systematic synthesis." *Proceedings of 2nd Conference on Electronic Computation*, ASCE, 105-122.

Schueller, G.I. (2007). "On the treatment of uncertainties in structural mechanics and analysis." *Comupters and Structures*, 85, 235-243.

Senthooran, S., Lee, D.D., and Parameswaran, S. (2004). "A computational model to calculate the flow-induced pressure fluctuations on buildings." *Journal of Wind Engineering and Industrial Aerodynamics*, 92, 1131–1145.

Shapiro, L.J. (1983). "The asymmetric boundary layer flow under a translating hurricane." *Journal of Atmospheric Sciences*, 40(8), 1984-1988.

Shinozuka, M. (1983). "Basic analysis of structural safety." *Journal of Structural Engineering*, 109, 721-740.

Shinozuka, M., and Jan, C.M. (1972). "Digital simulation of random processes and its applications." *Journal of Sound and Vibration*, 25(1), 111-128.

Shinozuka, M., and Deodatis, G. (1997). "Simulation of stochastic processes and fields." Probabilistic Engineering Mechanics, 12(4), 203–207.

Simiu, E., and Heckert, N.A. (1998). "Ultimate wind loads and directional effects in non-hurricane and hurricane-prone regions." *Environmetrics*, 9, 433-444.

Simiu, E., and Scanlan, R.H. (1996). Wind effects on structures: fundamentals and applications to design. John Wiley, New York.

Smith, B.S., Coull, A. (1991). Tall building structures: analysis and design. John Wiley & Sons, INC.

Spencer Jr BF, Nagarajaiah S (2003). "State of the art of structural control." *Journal of Structural Engineering*, ASCE, 129(7), 845–56.

Socquet-Juglard H., Dysthe, K., Trulsen, K., Krogstad, H., and Liu, J. (2005). "Probability distributions of surface gravity waves during spectral changes." *Journal of Fluid Mechanics*, 542, 195-216.

Solari, G. (1996). "Evaluation and role of damping and periods for the calculation of structural response under wind loads." *Journal of Wind Engineering and Industrial Aerodynamics*, 59, 191-210.

Solari, G. (1997). "Wind-excited response of structures with uncertain parameters." *Probabilistic Engineering Mechanics*, 12(2), 75-87.

Solari, G. (2002). "The role of analytical methods for evaluating the wind-induced response of tructures." *Journal of Wind Engineering and Industrial Aerodynamics*, 90, 1453-1477.

Song, J., and Der Kiureghian, A. (2006). "Joint First-Passage Probability and Reliability of Systems under Stochastic Excitation." *Journal of Engineering Mechancis*, 132(1), 65–77.

Sousa LG, Cardoso JB, Valido AJ (1997). "Optimal cross-section and configuration design of elastic–plastic structures subject to dynamic cyclic loading." Structural Optimization, 13, 112–118.

Stathopoulos, T. (1997). "Computational wind engineering: Past achievements and future challenges." *Journal of Wind Engineering and Industrial Aerodynamics*. 67-68, 509-532.

Stathopoulos, T. (2002). "The numerical wind tunnel for industrial aerodynamics: real or virtual in the new millennium?" *Wind and Structures*, 5(2-4), 193-208.

Surahman A. and Rojiani K.B. (1983). "Reliability based optimum design of concrete frames." *Journal of Structural Engineering*, 109(3), 741-757.

Tallin, A., and Ellingwood, B. (1985). "Wind induced lateral-torsional motion of buildings." *Journal of Structural Engineering*, 111(10), 2197-2213.

Tamura, Y., K. Shimada and H. Yokota (1994). "Estimation of structural damping of buildings." Proc. ASCE Structural Congress and IASS Int, Symp., Atlanta, USA, 2, 1012-1017.

Tamura, Y., Suganuma, S. (1996). "Evaluation of amplitude-dependent damping and natural frequency of buildings during strong winds." *Journal of Wind Engineering and Industrial Aerodynamics*, 59, 115-130.

Tamura, Y., Kawana S., Nakamura, J., Kanda, J., and Nakata, S. (2006). "Evaluation perception of wind-induced vibration in buildings." *Proceedings of the Institution of Civil Engineers: Structures & Buildings* 159, 283-293.

Truman, K.Z., and Cheng, F.Y. (1997). "How to optimize for earthquake loads." Guide to Structural Optimization, ASCE Manuals and Reports on Engineering Practice No. 90, 237-261.

Tschanz, T. and Davenport, A. G. (1983). "The base balance technique for the determination of dynamic wind loads." *Journal of Wind Engineering and Industrial Aerodynamics*, 13, 429–439.

Tse, T., Kwok, K. C. S., Hitchcock, P. A., Samali, B., and Huang, M. F. (2007). "Vibration control of a wind-excited benchmark tall building with complex lateral-torsional modes of vibration." *Advances in Structural Engineering*, 10 (3), 283-304.

Tu J, Choi KK (1999). "A new study on reliability-based design optimization." *Journal of Mechanical Design*, 121(4), 557–564.

Tvedt L. (1990). "Distribution of quadratic forms in normal space-application to structural reliability." *Journal of Engineering Mechanics*, 116(6), 1183–1197.

Van Der Hoven. (1957). "Power spectrum of horizontal wind speed in the frequency range from 0.0007 to 900 cycles per hour." *Journal of Meteorology*, AMS, 14, 160.

Vanmarcke E. H. (1972). "Properties of spectral moments with applications to random vibration."

*Journal of Engineering Mechanics*, 98 (EM2), 425-446.

Vanmarcke E.H. (1975). "On the distribution of the First-passage time for normal stationary random processes." *Journal of Applied Mechanics*, 42, 215-220.

Venkayya, V.B., Khot, N.S., and Reddy, V.S. (1968). "Optimization of structures based on the study of strain energy distribution." AFFDL-TR-68-150.

Vickery, B.J. (1966). "On the assessment of wind effects on elastic structures." *Australian Civil Engineering Transactions*, CE8, 183–192.

Vickery, B.J., Isyumov, N., and Davenport, A.G. (1983). "The role of damping, mass and stiffness in the reduction of wind effects on structures." *Journal of Wind Engineering and Industrial Aerodynamics* 11, 285–294.

Vickery, B. J. and Daly, A. (1984). "Wind tunnel modelling as a means of predicting the response to vortex shedding." *Engineering Structures*, 6, 363–368.

Vickery P.J., Twisdale L.A. (1995a). "Prediction of hurricane wind speeds in the United States." *Journal of Structural Engineering*, 121 (11) 1691-1699.

Vickery P.J., Twisdale L.A. (1995b). "Wind-field and filling models for hurricane wind-speed prediction." *Journal of Structural Engineering*, 121 (11) 1691-1699.

Vickery P.J., Skerlj P.F., Steckley A.C., and Twisdale L.A. (2000). "Hurricane wind field model for use in hurricane simulations." *Journal of Structural Engineering*, 126 (10) 1203-1221.

Von Karman, T. (1948). "Progress in the statistical theory of turbulence." *Proceedings of the National Academy of Sciences*, Washington, DC, 530-539.

Vrouwenveldera, A.C.W.M. (2002). "Development towards full probabilistic design codes." *Structural Safety*, 24, 417–432.

Wang BP (1991). "Improved approximate methods for computing eigenvector derivatives in structural dynamics." *AIAA Journal*, 29, 1018–1020.

Wen, Y.K. (2001). "Reliability and performance-based design." *Structural Safety*, 23, 407-428.

Wilson, E.L., Der Kiureghian, A., and Bayo, E.P. (1981). "A replacement for the SRSS method in seismic analysis." *Earthquake Engineering and Structural Dynamics*, 9, 187-192.

Wu WF, Lin YK. (1984). "Cumulant-neglect closure for non-linear oscillators under random parametric and external excitations." *International Journal of Non-Linear Mechanics*, 19(4), 349–62.

Xie, J., Kumar, S., and Gamble, S. (2003). "Wind loading study for tall buildings with similar dynamic properties in orthogonal directions." *Proceedings of the 11th International Conference on Wind Engineering*, 2-5 June 2003, Texas Tech University, 2390-2396.

Xu, L., Gong, Y.L., and Grierson, D.E. (2006). "Seismic Design Optimization of Steel Building Frameworks." *Journal of Structural Engineering*, 132(2), 277-286.

Xu, Y.L., Zhan, S. and Ko, J.M. (2000). "Effects of Typhoon Sam on Di Wang Tower: Field measurement." *HKIE Transactions*, 7(2), 41-48.

Xu YL, Chen B. (2007a). "Integrated vibration control and health monitoring of building structures using semi-active friction dampers: Part I — Theory." *Engineering Structures* (In press).

Xu YL, Chen B. (2007b). "Integrated vibration control and health monitoring of building structures using semi-active friction dampers: Part II —Numerical investigation." *Engineering Structures* (In press).

Yang, C. Y. (1985). Random vibration of structures. Wiley, New York.

Yip, D. Y. N., and Flay, R. G. J. (1995). "A new force balance data analysis method for wind response predictions of tall buildings." *Journal of Wind Engineering and Industrial Aerodynamics*, 54/55, 457-471.

Youn, B.D., Choi, K.K., and Park, Y.H. (2003). "Hybrid Analysis Method for Reliability-Based Design Optimization." *Journal of Mechanical Design*, 125, 221-232.

Youn, B.D., Choi, K.K., and Du, L. (2005). "Adaptive probability analysis using an enhanced hybrid mean value method." *Structural and Multidisciplinary Optimization*, 29, 134-148.

Zhang DL, Liu Y, and Yau MK. (2000). "A multi-scale numerical study of hurricane Andrew, 1992: Part III: Dynamically induced vertical motion." *Monthly Weather Review*, 128, 3772-3788.

Zhou, Y., Gu, M., and Xiang, H. F. (1999). "Along-wind static equivalent wind loads and response of tall buildings. I: Unfavorable distributions of static equivalent wind loads." *Journal of Wind Engineering and Industrial Aerodynamics*, 79(1–2), 135–150.

Zhou Y, Kijewski T, Kareem A. (2003). "Aerodynamic loads on tall buildings: An interactive database." *Journal of Structural Engineering*, ASCE, 129, 394-404.

Zhu, W.Q. (1990). "The exact stationary response solution of several classes of nonlinear systems to white noise parametric and/or external excitations." *Applied Mathematics and mechanics*, 11(2), 165-175.

Zhu W.Q., Cai G.Q., and Lin Y.K. (1990). "On exact stationary solutions of stochastically perturbed Hamiltonian systems." *Probabilistic engineering mechanics*, 5(2), 84-87.

Zhu W.Q., Yang Y.Q. (1996). "Exact stationary solution of stochastically excited and dissipated Hamiltonian systems." *Journal of Applied Mechanics*, ASME, 63, 493-500.

Zhu W.Q., Huang Z.L. (2001). "Exact stationary solutions of stochastically excited and dissipated partially integrable Hamiltonian systems." *International Journal of Non-Linear Mechanics*, 36, 39-48.

Zou, T., and Mahadevan, S. (2006). "A direct decoupling approach for efficient reliability-based design optimization." *Structural and Multidisciplinary Optimization*, 31, 190-200.

Zou, X.K., and Chan, C.M. (2005). "An optimal resizing technique for seismic drift design of concrete buildings subjected to response spectrum and time history loadings". *Computers and Structures*, 83(19-20), 1689-1704.

Zuranski, J. A., Jaspinska, B. (1996). "Directional analysis of extreme wind speeds in Poland." *Journal of Wind engineering and Industrial Aerodynamics*, 65, 13-20.

* 9 7 8 3 6 3 9 1 6 2 5 3 0 *